Thinking
Statistically

Thinking Statistically

Anthony Banfield

Library of Congress Control Number:		2017903995
ISBN:	Hardcover	978-1-5245-9200-4
	Softcover	978-1-5245-9199-1
	eBook	978-1-5434-1214-7

Print information available on the last page.

Rev. date: 08/21/2019

To order additional copies of this book, contact:
Xlibris
1-888-795-4274
www.Xlibris.com
Orders@Xlibris.com
755451

CONTENTS

PREFACE

In everyday life, persons (and that includes you who are reading this text) either consciously or unconsciously ask questions or make statements that are statistical in nature. Questions such as, How many points did a particular basketball player score in the last five games he played? How many goals did a particular soccer team concede in the last six games played? What are the heights of the thirty female students in a form 4 class? These are all related to statistics. Similarly, statements such as "On average, I get six hours of sleep each night" or "The average rainfall for the month of June was ten millimetres" or "For every one thousand persons who take a particular over-the-counter drug, about four persons experience slight nausea as a side effect" are all statistical in nature. In addition, when reading newspapers or magazine articles, information is frequently presented statistically in the form of tables or pictures (e.g., bar graphs, pie charts, histograms, and scatter plots).

Statistics deals with large amounts of numbers and may be formally described as that branch of mathematics that is concerned with the collection, organisation, summarisation, analysis, presentation, interpretation, and prediction of numerical data. The first step in any statistical analysis is always the collection of data in the form of numerical facts. Data may be collected on a variety of things, for example,

- People—their heights and/or weights
- Weather—number of days for the year it rained and the amount of rainfall for each of these days
- Sports—the amount of punches scored by each boxer in each round of a boxing match
- Drugs—the number of people who experienced a particular side effect after taking a drug
- Manufacturing—the amount of a particular product produced not conforming to specifications

As we can see, statistics surrounds our lives, and we must be in a position to at least understand questions and statements that are statistical in nature—that is, we must be able to think statistically. And so we have the title of this text: *Thinking Statistically*. In this elementary text, we will be discussing methods of analysing and presenting data, interpreting the data, and also using collected data to make predictions. There are numerous completely solved examples within the body of the text that is hoped would bring the theory into perspective and demonstrate to the reader in a clear manner the mechanics of solving problems that are statistical in nature. The book should be of value for anyone interested in getting started in statistics and wanting to get a feel of what statistics is all about. It should also be of value for anyone who needs a refresher in the basics of statistics. This book can be used in conjunction with any course where statistics is being introduced and will also be of particular value for private study as it contains numerous solved problems. The book has been written for you. I hope you enjoy reading it as much as I have enjoyed writing it. Any comments about the text are welcome and can be sent directly to the author at the following e-mail address: apolloahrb@ gmail.com. Thank you.

CHAPTER 1

Understanding Data and Its Content

1.1 What Is Data?

The word *data* is actually a Latin word meaning "those that are given." The singular form is *datum*, meaning "that which is given." Data can be thought of as information received or collected. The directions that someone gives you on how to get to a particular place is a datum. The number of runs scored by a batsman in a particular inning is a datum. In the two examples given, the singular form, *datum*, is used because in either case, a single amount of information is received or collected.

For information received or collected to be subjected to statistical analysis, a multiple amount of information (data) must be collected on a given *object* or *event*. To clarify, suppose a teacher wants to investigate the heights of the twenty-five students in her class. In order to do this, she attaches a three-metre (300 cm) tape to the wall with the zero end of the tape coincident with the floor. She then asks each student in turn to stand in front of the tape (shoes removed) and records the measured heights. In this case, the *object* under investigation is each student in the class, and the twenty-five measurements obtained are referred to as a **data set**, on which statistical analysis can be performed. Suppose the same teacher

1

now asks one of her students to roll a die and record the value that comes up. The *event* is rolling the die, and the result (measurement) of the event is the number on the die that comes up. If the teacher now asks each of the twenty-five students in the class to roll the die and record which number comes up, the result is twenty-five observations being recorded instead of one, as in the previous case. In the latter case, we can again apply statistical analysis to the recorded data set. Note, in this die-rolling experiment, the die is the *object* under investigation.

The teacher now decides to carry out an investigation as to the type of transmission each car has that is parked in the school's car parking lot. To do this, each car is examined, and a record is made as to whether the car has automatic or manual transmission. In this case, the cars are the objects under investigation, but unlike the above examples, the data recorded is not a number (i.e., numerical value) but a description, namely, automatic transmission or manual transmission.

Observations made on an object or event may be recorded either as a numerical value (in which case the observation is called a measurement or a score), or it may be recorded as a nonnumerical value (in which case the observation is descriptive).

1.2 Types of Data

Data collected by carrying out measurements on some numerical scale are called **quantitative data**, whereas data collected in nonnumerical form (descriptive) are called **qualitative data**. Examples of quantitative data are:

- the heights of students in a class measured in centimetres;
- the weights of students in a class measured in kilograms; and
- the test score of each student in a class who wrote a mathematics exam.

Qualitative data are descriptive data that can only be categorised or grouped into **classes**. Some examples follow:

- The transmission type (manual or automatic) of the cars parked in a parking lot
    In this case, the investigation is to collect information on the type of transmission each car has. The two categories into which a car may be placed are either "automatic transmission" or "manual transmission."

- The gender (male or female) of persons attending a cinema show
    In this case, the investigation is to determine the number of males and females attending a cinema show.

- The type of cellular phone (Nokia, Motorola, LG, Samsung) the staff of an office have
    In this case, the cellular phones owned by the office staff can be placed into one of the four above categories.

- The degree status of a group of university graduates
    The degree status is a qualitative variable, and the different categories or classes may be BSc, MSc, and PhD.

- The ranking of a group of military officers
    The ranking is a qualitative variable, and the categories or classes may be corporal, sergeant, lieutenant, captain, major, colonel, or general.

Whether the event that is observed is quantitative (expressed as numbers) or qualitative (a description), a record can be made of the number of times each event occurs. The *number of times* each event occurs is called the **frequency** of the event; for example, the number of students who receive a particular score in a biology exam or the number of cars in the parking lot that have manual transmission.

1.3 Raw Data

When data is collected about an object or event and is recorded in no particular order or pattern, they are called **raw data**.

Consider a class of 25 male students whose weights are to be measured. Students are called one at a time to stand on a scale and have their weight recorded. The recorded weights are shown in table 1.1 below.

Table 1.1 Weights in kg of 25 Male Students

63	68	60	57	71
66	64	67	67	65
76	63	80	54	75
66	64	67	69	61
73	61	64	71	58

The weights listed in table 1.1 above are in no particular order or pattern and is therefore referred to as raw data.

In the jargon of statisticians, the students (whose weights are being measured) are the objects under investigation and are referred to as **experimental units**, and the weight of each student, which is the property or characteristic being measured and which varies from student to student, is called the **variable**. The variable about which information is collected and recorded may be assigned either numerical values or descriptions. If the variable is assigned numerical values, it is called a **quantitative variable**, but if it is assigned descriptions, it is called a **qualitative variable**. For example, when investigating the weights of students in a class, the variable—weight—is assigned numerical values and hence described as a quantitative variable. But if one is investigating the brand of shoes worn by students in a gym class, the variable—brand—is assigned descriptive categories (e.g., Adidas, Nike, Reebok, Puma) and is called a qualitative variable.

When examining a data set for the first time, one must clearly ascertain the following:

- What is the experimental unit being investigated?
- What is the variable on which data is collected?
- Is the data collected qualitative or quantitative?
- If the data is qualitative, what are the categories or classes into which it is divided?

Table 1.2 below gives examples of the features of some data sets.

Table 1.2 Features of Some Data Sets

Data set	Experimental unit	Variable	Type of data		If Qualitative, state classes
			Quantitative	Qualitative	
Height of students	Students	Height	Yes	---------	-----------
Gender of persons attending a cinema show	Persons	Gender	----------	Yes	Male, Female
Years imprisonment for persons committing a crime	Prisoners	Years imprisonment	Yes	----------	---------
Heads obtained from flipping 4 coins 25 times	The four coins	No. of heads	Yes	-----------	-----------
Movies preferred (action, horror, suspense, drama) by 100 moviegoers	100 moviegoers	Type of movie	-----------	Yes	Action, Horror, Suspense, Drama
Corrosiveness of 12 chemicals	The 12 chemicals	Corrosiveness	----------	Yes	Corrosive, Noncorrosive
Households in a particular area that use metal or PVC water pipes	Households	Type of pipe	-------------	Yes	Metal, PVC

1.4 Population or Sample

As we have seen, a data set contains information about some variable of an object or event that is of interest to us. If the data set contains information about *all* of the objects (person or thing) or events, then we say the data set describes the **population**. However, if the data set contains information for only a *portion* of all the objects or events under consideration, then the data set is for a **sample** of the population. Consider the following examples:

1. If it is desired to know the weights of *all* 25 students in a class, we measure and record the weight of each of the 25 students. The listing of the 25 weights make up the data set for the entire class and is called the population data set because the target of interest are the weights of *all* the students in the class. On the other hand, had we recorded the weights of only 10 students in the class, then the data set, now consisting of 10 weights, would be for a sample of the population.

2. Suppose we are interested in the length of cucumbers in a newly harvested cucumber field. The data set describing the population would be the length of *all* newly harvested cucumbers. A sample data set would be, say, the length of 100 newly harvested cucumbers.

3. Suppose a retailer of light bulbs buys a new shipment of 2000 bulbs and the manufacturer of the bulbs claim that the average lighting life of a bulb before it fails is 2500 hours. This claim may be tested by the retailer by lighting *all* the bulbs from the shipment (the population) continuously until they fail to determine the lighting life in hours. However, by doing this, all of the shipment would be destroyed and so is not practical, as no bulbs would be left over for resale. The manufacturer's claim may still be examined if a *portion* (sample) of say, thirty randomly selected bulbs were checked.

In many instances, it may be too costly, too time-consuming, or just impractical to measure all the experimental units in a population. In such cases, a portion (sample) of the experimental units is selected for measurement from the population. The results obtained from examining the sample may then be used to make predictions about the entire population. To illustrate the concept, suppose 20 of the 30 randomly selected light bulbs failed before the 2500 hours were up, then it can be shown using statistics that most of the light bulbs in the shipment does not meet the manufacturer's claim.

1.5 Random Sample

When information is required about a population and the population is very large, a sample from the population is chosen, and information about the sample is determined, which is then used to make inferences (predictions) about the population. For the inferences to be accurate, the sample chosen must be **representative** of the population, meaning the sample chosen must possess characteristics typical of those possessed by the target population. For the sample to be representative of the population, the sample chosen must be a **random sample**. Randomness means that the sample size chosen is chosen in such a way that any sample of the same size has an equal chance of being selected.

Manual random number generation can be used if the population is not too large. In such a case, each observation is recorded on a piece of paper and placed in a container. The container is then closed and shaken to mix the pieces of paper thoroughly. The lid is then removed and the number of pieces of paper corresponding to the required sample size is then taken out. The observations recorded on the removed pieces of paper are the ones making up the random sample.

Example 1.1

State whether each of the following variables is qualitative or quantitative.

a) The amount of students in a class
b) The number of CDs sold by a music store on a given day
c) The brand of jeans worn by the members of a class
d) The school level of the children in a household
e) The height achieved by a pole-vaulter
f) The variety of tropical fruits grown on an estate
g) The flavours of jams sold at a supermarket

Solution 1.1

a) Quantitative. The amount of students is measured as a numerical quantity.
b) Quantitative. The number of CDs sold is reported as a numerical value.
c) Qualitative. Brand cannot be reported as a numerical value but can only be classified into categories, e.g., Wrangler, Lee, Levi's, Calvin Klein, Jordache.
d) Qualitative. School level may be categorized as kindergarten, elementary, secondary, or university. These are nonnumerical values.
e) Quantitative. Height is measured on a numerical scale, e.g., metres.
f) Qualitative. Fruit varieties may be classified as mangoes, portugals, oranges, grapefruit, or passion fruit. These are non numerical quantities.
g) Qualitative. Jam flavours are nonnumerical values and may be classified as strawberry, pineapple, peach, guava, or apricot.

Example 1.2

A marketing representative for an ice cream company is interested in finding out which flavours of ice cream are most liked by children between the ages of 7 and 10 years in a particular region. To do this, the marketing representative visited three randomly selected elementary schools in the particular region and asked each of the children which flavour of ice cream he/she liked best and recorded the results. A total of 500 children were interviewed, which made up the total enrollment of the three schools.

a) What data (information) is being collected by the marketing representative?
b) Are the data collected qualitative or quantitative?
c) What is the population of interest to the marketing representative?
d) Describe the sample.
e) What are the experimental units?

Solution 1.2

a) The flavours of ice cream liked by children between the ages of 7 and 10 years
b) Qualitative. Ice cream flavours are not recorded as numerical values but as categories, e.g., vanilla, strawberry, chocolate, pistachio, etc.
c) *All* children between the ages of 7 and 10 years in the particular region.
d) The 500 children between 7 and 10 years of age who were interviewed
e) Since the data is being collected from the children between 7 and 10 years old, the children are the experimental units.

Example 1.3

A computer salesman is interested in determining how many households in a small town of 800 households have a computer. He carries out his investigation by randomly selecting 150 households, which he subsequently visits to ask if there is a computer in the home. His results showed that of the 150 households polled, 35 of these (or 23%) had a computer.

a) Identify the experimental units in this investigation.
b) Describe the variable measured on each experimental unit and state its type.
c) Describe the target population.
d) Describe the sample.
e) What percentage of households from the sample have a computer?

Solution 1.3

a) The households are the experimental units. Recall that the experimental unit is the object about which information is measured or collected. Here, we are collecting information about whether or not a household has a computer. Thus, each household is the experimental unit.
b) The variable measured is the computer status, meaning the presence or absence of a computer for the 150 households selected from the small town. Since the measurement may be recorded as "household has a computer" or "household does not have a computer," which are both nonnumerical values, the variable measured is qualitative.
c) The target population is the computer status of *all* 800 households in the small town.
d) The sample is the 150 households interviewed by the computer salesman.
e) 23%

Statistical Terms

data set. A collection of information about an object or event

quantitative data. Observations measured on some numerical scale

qualitative data. Observations recorded as a descriptive category

class. A descriptive or numeric category into which collected data can be placed

frequency. The number of times an event occurs

raw data. Information collected and recorded in no particular order about some variable related to an object or event under investigation

experimental unit. The object or event about which information (data) is collected

variable. The property or characteristic that varies from one observation to the next

quantitative variable. A variable that is assigned numerical values

qualitative variable. A variable that is assigned descriptive categories

population. The entire set of items under investigation

sample. A portion of the data selected from a population

representative sample. A sample of data that exhibits characteristics typical to those possessed by the population from which the sample was taken

random sample. A sample of a given size selected from a population such that any sample of that same size has an equal chance of being selected

CHAPTER 2

A Picture Saves a Thousand Words: Bar Graphs and Charts

2.1 Why Represent Data as a Picture?

As was mentioned in chapter 1, when data are originally collected, it is in no particular order or pattern, and as a result, the meaning and interpretation of the recorded data can be difficult to grasp. To bring some clarity to this raw data, it must first be organised and summarised before it can be presented as a picture. Common methods of presenting data pictorially are **bar graphs** and **pie charts**. When data is presented in these picture forms, it makes it easier to understand and interpret.

2.2 Presenting Data in the Form of Bar Graphs

A bar graph is a two-dimensional plot of a numerical vs. nonnumerical pair. That is, one axis of the bar graph (either the x-axis or the y-axis) has a scale of numerical values, and the other axis has no numerical scale. The bars are drawn having the same width but may be oriented either vertically or horizontally. In either case, the height or length of the bars are proportional to the numerical quantities in question.

2.2.1 Vertical Bar Graphs (Column Charts)

Consider the hypothetical rainfall figures by month for a two-year period over the years 2001 and 2002 as shown in table 2.1. This table is a data set that gives information about the amount of rainfall by month for a region for the years 2001 and 2002. What is the experimental unit? What is the variable being observed? Is the variable qualitative or quantitative?

Since the amount of rainfall is being observed on a monthly basis, the experimental unit, which is the object under investigation, is the rainfall. The variable being observed (measured) is the *amount* of rainfall measured in millimetres. The variable is quantitative as it is measured against a linear numerical scale.

Table 2.1 Monthly Rainfall Figures in mm for a Region

	2001	2002
Jan	137.4	111.8
Feb	7.81	56.4
Mar	122.6	39.9
Apr	79.2	44.4
May	36.3	35.8
Jun	88.9	88.9
Jul	91.1	91.2
Aug	91.7	91.7
Sep	2.8	92.2
Oct	211.8	211.8
Nov	82.6	124.7
Dec	202.6	121.9

Although the rainfall figures over the two-year period are summarised in tabular form, it is still somewhat difficult to recognise a pattern easily. Consider these questions:

1) Which are the wettest and driest months?
2) Which seems to be the driest part of the year?
3) What month has the highest chance of flooding?

Now consider the same information given in table 2.1, presented as a bar graph as shown in figure 2.1, with months along the horizontal axis and the amount of rainfall along the vertical axis. Note that for one axis, months are assigned, which is not a numerical quantity.

Fig. 2.1 Monthly Rainfall for 200′

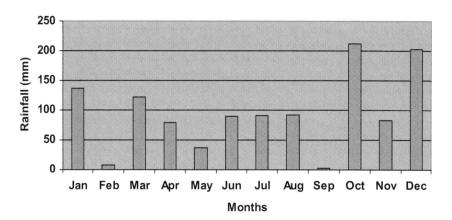

In the bar graph, the information (in this case, the monthly rainfall figures) is represented as a series of vertical bars, all of the same width. The height or length of each bar represents the amount of rainfall for each month during the year. I am sure you will agree that the bar graph presentation of the tabular data is easier to interpret, and the trend or variation of the rainfall throughout the year is more easily grasped.

Figure 2.2 shows the bar graph presentation for the rainfall figures for 2002.

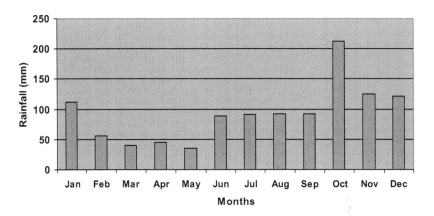

Fig. 2.2 Monthly Rainfall for 2002

Using graph paper, the bar graph may be constructed as follows:

1) On the longer side of the sheet, set up the horizontal x-axis by taking 1 cm of width for each month, leaving 1-cm gaps between each month.
2) For the vertical y-axis, take 1 cm to represent 20 mm of rainfall.
3) Plot the rainfall figures (data) for each of the months.

By plotting the rainfall figures for corresponding months over the two-year period, a comparative (side-by-side) bar graph is obtained (fig. 2.3). This allows the easy comparison of rainfall figures for similar months over the two-year period.

From figure 2.3, it can be seen that the driest month for 2001 was September, whereas for 2002, it was May. The wettest month for both years was October, and the chances of flooding would have been the greatest during that month. However, for 2001, December was also very wet, with a high chance of flooding. The comparison also shows that June, July, and August for both years had almost equal amounts of rainfall.

**Fig.2.3 Comparative Bar Chart of Rainfall
for 2001 and 2002**

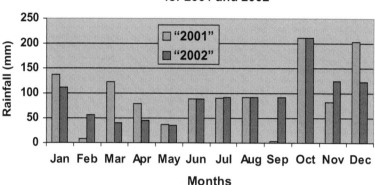

2.2.2 Horizontal Bar Graphs

Bar graphs may also be plotted using horizontal bars. Consider table 2.2 below, which shows the mode of transportation of the students in a form 3 class. Examination of the data set reveals that the experimental units are the students, the variable of interest is the mode of transportation—which is a qualitative variable—and the classes or categories of the qualitative variable are cycling, walking, public transport, and private car.

Table 2.2 Mode of Transportation of Students in a Form 3 Class

Mode of transportation	Number of students
Cycling	3
Walking	5
Public transport	12
Private car	10

From table 2.2, it can be seen that three students cycle to school. Alternatively, it can be said that the number of observations falling within the cycling category (or class) is 3. Similarly, it can be said that the number of observations falling within the walking, public transport, and private car categories are 5, 12, and 10 respectively.

In the jargon of the statistician, it is said that the frequency of students in the cycling category is 3. The frequency of the students in the walking category is 5. The frequency of the students in the public transport category or the frequency of the students who come to school using public transport is 12, and the frequency of the students falling within the private car category (or class) is 10.

The number of observations falling within a particular category or class is called the **class frequency**. Thus, the cycling class frequency is 3, the walking class frequency is 5, the public transport class frequency is 12, and the private car class frequency is 10.

The horizontal bar graph for the information in table 2.2 is presented below in figure 2.4.

Fig. 2.4 Students Mode of Transportation

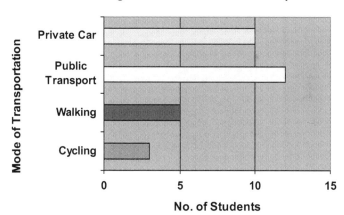

2.3 Presenting Data in the Form of a Pie Chart

A pie chart is also used to present information when one of the variables are nonnumerical. Each category or class of the qualitative variable is represented by a slice (sector) of a pie, the size of each slice (angle of the sector measured in degrees) being proportional to the quantities in question. The information in table 2.2 showing the modes of transportation of the form 3 students are shown on a pie chart in figure 2.5.

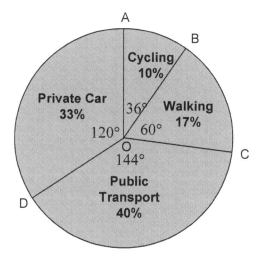

Fig, 2.5 Students Mode of Transportation

The mode of transport is a qualitative variable, and each slice (sector) of the pie represents one of the categories, i.e., cycling, private car, walking, and public transport. It can be seen from the chart that 10% of the students in the form 3 class ride to school, 17% walk to school, 33% are driven by private car, and 40% use public transport, which may be bus, maxi taxi, or regular taxi. The format of the calculations is summarized in table 2.3 below.

Table 2.3 Format of Calculations for Drawing a Pie Chart

Mode of transportation	Number of students (Frequency)	Fraction or proportion of students (Relative frequency)	Angle of sector (Degrees)	Percentage
Cycling	3	3/30=0.100	0.100×360=36	0.100×100=10.0
Walking	5	5/30=0.166	0.166×360=60	0.166×100=16.6
Public transport	12	12/30=0.400	0.400×360=144	0.400×100=40.0
Private car	10	10/30=0.333	0.333×360=120	0.333×100=33.3
TOTAL	30			

The pie chart is drawn as follows:

1. First, calculate the fraction or proportion (relative frequency) of the students who use a particular mode of transport.
2. This fraction or proportion of students for each mode of transportation is then multiplied by 360 degrees (because a circle has 360 degrees) to determine the angle for each sector (i.e., class or category) of the pie.
3. Using a compass, a circle of convenient radius is drawn, e.g., 3 or 4 cm.
4. Draw a vertical line from the centre of the circle to the top of the circle (from point O to A). This is the reference or starting line from which the first sector is marked off. The first sector may be drawn either to the right or to the left of the reference line.
5. Position the centre of the protractor at the centre of the circle with the 0–180 degree line over the vertical line (OA) drawn and draw the line OB at 36° to the right of OA. The sector OAB formed defines the cycling category, which also represents 10% of the students in the class.
6. Using the protractor in a similar manner, the line OC is marked off at 60° from the line OB. The sector OBC depicts

the walking category (i.e., the students who walk to school) and represents 17% of the students in the class.
7. In a similar manner, the remainder of the pie chart is completed.

It can be seen that given the same information, it is easier to construct a bar graph than a pie chart because it is easier to calculate the lengths of bars than to calculate the angles of sectors. Also, the brain can more easily compare the lengths of bars than the angular sizes or areas of sectors. As a result, the bar graph is better for general use than the pie chart, especially when the sizes of the various parts are compared to one another. When, however, the sizes of the various parts are to be compared with the whole, the pie chart presentation is more desirable.

Statistical Terms

bar graph. A diagrammatic form of presenting information about two variables where one variable is descriptive and the other variable is numerical. The width of the bars represent the descriptive variable, and all are made the same width, and the height or length of the bars are proportional to the numerical quantities.

pie chart. A diagrammatic form of presenting information as the sectors of a pie, where each sector represents the descriptive category of the data and the angle of the sector is proportional to the corresponding numerical quantity.

class frequency. The number of observations falling into a given class.

relative frequency. The number of times an observation occurs, expressed as a fraction or proportion of the total number of observations.

CHAPTER 3

Keeping Count of Event Occurrences (Frequency Distributions)

3.1 Organising and Summarising Raw Data

Recall that raw data are collected data that have not been organised into any particular order. Before raw data can be used meaningfully for analysis and interpretation, it must be organised and summarised as it is only then that it is possible to present the information (data) in a convenient form, thereby revealing any patterns.

The first step in organizing raw data is to summarise the data as a table called a **frequency distribution table**. The term frequency was already mentioned and defined as the number of times an event occurs or the number of observations falling into each class. A frequency distribution table is a tabular summary and representation of collected data by class together with the number of observations associated with (or falling into) each class. That is, the table is a pairing of class with corresponding class frequency. The following example illustrates the procedure for transforming a raw data set into a frequency distribution table, also referred to as a **frequency table**.

Example 1

Suppose we are interested in determining the graduating degree status of a sample of 25 graduating students from UWI St. Augustine. The students making up the sample were asked what type of degree they obtained. The degree status, which is a qualitative variable, was recorded as BSc, MSc, and PhD. Table 3.1 shows the degree status of the 25 graduating students.

The task at hand is to construct a frequency table from this raw data. Notice that the raw data looks disordered. The first step is to summarise the data into the three classes or categories of BSc, MSc, and PhD in order to determine the number of observations falling into each class, called the **class frequency**. The required frequency distribution is shown in table 3.2.

Table 3.1 Degree Status of 25 Graduates

Name	Degree tatus	Name	Degree status
Emma	MSc	Svetlana	BSc
John	BSc	Sandra	MSc
Dexter	BSc	Maribel	PhD
Lisa	BSc	Joseph	BSc
David	MSc	Brian	BSc
Marc	PhD	Shara	BSc
Victoria	BSc	Anna	BSC
Yulia	MSc	Gregory	BSc
Deborah	PhD	Roger	PhD
Adriana	BSc	Luke	MSc
Paul	MSc	Michael	BSc
Colin	BSc	Carolina	MSc
Yana	BSc		

Table 3.2 Frequency Table for Degree Status Data

Degree status (Class)	Tally	No. of students (Frequency)
BSc	⊮ ⊮ ////	14
MSc	⊮ //	7
PhD	////	4
Total		25

Referring to table 3.2, the centre column is called a tally sheet and is used to tabulate the number of observations for each class from the raw data. The tallying is done by looking through table 3.1 row by row and recording a tally mark (stroke) for each degree status that occurs. This is recorded in table 3.2 in the second column of the appropriate row. The fifth stroke is drawn through the four previously recorded strokes to form a grouping of five strokes. A count of the tally groups is then recorded in the 3rd column of table 3.2 for each of the three classes as the number of students graduating with BSc, MSc, and PhD status respectively, called the class frequency. This tally sheet is omitted in the final presentation of the frequency distribution table.

If the class column in a frequency distribution table contains descriptive information (i.e., nonnumerical data for the variable being observed), the frequency distribution table is referred to as a **qualitative frequency distribution table**. However, if the class column contains numerical values for the variable being measured, the frequency distribution table is referred to as a **quantitative frequency distribution table**.

If the variable being measured can only take on numerical quantities of integral values, the variable is called a **discrete variable**, whereas if the variable can have any value between two given values, it is called a **continuous variable**.

3.2 Some Frequency Distribution Tables

The class column of a frequency distribution table may be descriptive if the data is for a qualitative variable as shown in the preceding example (see also table 3.3), or it may be numerical in the case of quantitative variables. For quantitative variables, the class may be

(i) single-value numerical quantities (see table 3.4 and table 3.5), or
(ii) grouped numerical quantities (see table 3.6) for large data sets.

Grouping—although convenient and makes the data more manageable when dealing with large amounts of collected data— causes much of the detail in the original data to be lost but enhances the clarity of the overall picture of the data, thereby revealing important relationships. We will now take a look at some different frequency distribution tables.

Table 3.3 below shows a qualitative frequency distribution table for ice cream flavour preferences for a sample of 30 students at a secondary school.

Table 3.3 Qualitative Frequency Distribution
for Ice Cream Flavours Preferred
by 30 Students at a Secondary School

Ice cream flavours (Class)	No. of students liking (Frequency)	Relative frequency	Angle of sector for each slice (Degrees)
Vanilla	7	7/30 = 0.233	7/30 × 360 = 84
Strawberry	6	6/30 = 0.200	6/30 × 360 = 72
Chocolate	9	9/30 = 0.300	9/30 × 360 = 108
Cherry vanilla	4	4/30 = 0.133	4/30 × 360 = 48
Pistachio	3	3/30 = 0.100	3/30 × 360 = 36
Coconut	1	1/30 = 0.033	1/30 × 360 = 12
Total	30		

The ice cream flavour is a qualitative variable having the various flavours: vanilla, strawberry, chocolate, cherry vanilla, pistachio, and coconut as the different classes. For each class, the number of students who likes a particular flavour is listed. The table is interpreted as follows: It can be said that for vanilla ice cream, the frequency is 7, or the class frequency for vanilla ice cream is 7. In everyday language, we say that 7 of the 30 students prefer vanilla ice cream.

The calculations necessary to present the frequency distribution table as a pie chart are also illustrated in table 3.3. Column 3 of the table, called the **relative frequency**, is the statistical term for the fraction or proportion of students who likes a particular flavour of ice cream. This proportion is obtained by dividing the number of students who likes a particular flavour by the total number of students making up the sample. Column 4 of table 3.3 gives the calculation to determine the angle of the sector (measured in degrees) representing each flavour preferred. The sector angle size for each flavour is directly proportional to the number (or percentage) of students who likes that particular flavor.

Table 3.4 shows a quantitative frequency distribution table for adult male shoe sizes for 200 pairs of shoes sold by a particular shoe store.

Table 3.4 is a single-value discrete frequency distribution table. Although half sizes are present, the variable, which is the shoe size, is a discrete variable as in the shoe industry, half sizes are typical, in addition to full or integer sizes.

Table 3.4 Discrete Single-Value Frequency Distribution Table
for Shoe Sizes for 200 Pairs of Shoes Sold

Shoe sizes (Class)	No. of shoes sold (Frequency)
6	2
6 ½	5
7	12
7 ½	10
8	20
8 ½	15
9	45
9 ½	20
10	30
10 ½	12
11	14
11 ½	10
12	5
Total	200

Table 3.5 below shows a quantitative frequency distribution table for the weights of 50 cups of 175 ml ice cream measured to the nearest gram.

Table 3.5 Frequency Distribution for Weights of 50 Cups of 175 ml
Ice Cream Measured to the Nearest Gram

Weight of 175ml cups grams (Class)	No. of Cups (Frequency)	Relative Frequency	Angle of Sector for each slice (Degrees)
118	3	3/50 = 0.06	3/50 x 360 = 84
120	11	11/50 = 0.22	11/50 x 360 = 72
125	18	18/50 = 0.36	18/50 x 360 = 108
128	13	13/50 = 0.26	13/50 x 360 = 48
131	5	5/50 = 0.10	5/50 x 360 = 36
Total	50		

The weight is a quantitative variable with the following numerical classes: 118 g, 120 g 125 g, 128 g, and 131 g. Since the weight of each cup can be measured to varying degrees of accuracy depending on the resolution of the scale, it is quite possible to have measurements of 117.8 g, 118.5 g, 118.259 g, and 118.3345 g, all of which can be recorded as 118 grams to the nearest gram. Since the weight can take on any value depending on the resolution of the measuring instrument, it is regarded as a continuous variable. For example, the weight of 120 g indicates that the weight of a cup of ice cream can lie between 119.5 g and 120.5 g. Associated with each weight class is the number of cups having that weight when measured to the nearest gram. Thus, it can be said that for cups having a weight of 118 g, the frequency is 3, or the class frequency for cups weighing 118 g is 3. In everyday language, we say that 3 cups out of the sample of 50 cups weigh 118 g when measured to the nearest gram.

Table 3.6 shows a **grouped** frequency distribution table for the weights of 500 cups of 175 ml ice cream. The only difference between table 3.6 and table 3.5 is the amount of cups in the sample—500 in the former case and 50 in the latter. When the data set is large, as in this case, it is useful to group the measurement values into intervals and list the number of observations (measurements) falling into each interval. This reduces the actual amount of data to deal with, thereby making it more manageable.

Table 3.6 Grouped Frequency Distribution for Weights of 500 Cups of 175 ml Ice Cream

Weight of 175ml cups (Class)	No. of Cups (Frequency)	Relative Frequency	Angle of Sector for each slice (Degrees)
116 – 118	20	20/500 = 0.04	20/500 x 360 = 14.4
119 – 121	55	55/500 = 0.11	55/500 x 360 = 39.6
122 – 124	200	200/500 = 0.40	200/500 x 360 = 144.0
125 – 127	120	120/500 = 0.24	120/500 x 360 = 86.4
128 – 130	75	75/500 = 0.15	75/500 x 360 = 54.0
131 – 133	30	30/500 = 0.06	30/500 x 360 = 21.6
Total	500		

When the data set (collection of measurements) is large and contains repeated measurement values, it is practical to summarize the data set as a frequency distribution table. Such a table is formed by associating each measurement value with the number of times (frequency) that that measurement value occurs. The frequency table therefore has two columns, one with a heading that describes the measurement value and the other with the heading frequency, which states the number of times that particular measurement value occurs.

If the data set is extremely large, then the frequency distribution table can be made even more manageable by arranging the measurement values into intervals to form what is called a grouped frequency distribution table. Such a table is formed by arranging the measurement values into groups and associating each group of measurements with the number of measurements falling within each group. *It must be noted, however, that with a grouped frequency distribution table, we can only observe the number of measurements falling within a group (or measurement range) but have no idea as to the exact magnitude of the measurement values, thus losing some of the detail in the original data.*

When performing calculations with grouped frequency distribution tables, the midpoint value of each group is used as part of the calculation. This is discussed in section 3.3.

3.3 Components of a Grouped Frequency Distribution Table
3.3.1 Class Interval, Class Limits, and Class Mark

The grouping 116–118 in table 3.6 is called a **class interval** and is the first class interval of the six class intervals shown in the table. The end numbers, 116 and 118, are called the **class limits.** The smaller number, 116, is the **lower class limit**, and the larger number, 118, is the **upper class limit.** The midpoint of the class interval is called the **class mark** or the **class midpoint** and is determined by taking the arithmetic average of the lower and upper class limits. Thus, the class mark of the first class interval, 116–118, is (116+118)/2 = 117.

3.3.2 Class Boundaries

Referring again to table 3.6, the weights are recorded to the nearest gram, so the class interval 116–118 theoretically includes all weights from 115.5 g to 118.5 g and are called the class boundaries for the first class interval. The smaller number, 115.5, is the **lower class boundary**, and the larger number, 118.5, is the **upper class boundary**. For any grouped frequency distribution, the class boundaries for any class interval is found by adding the upper class limit of one class to the lower class limit of the next class and dividing by 2. For example, the class boundaries for the 4th class interval, 125–127, in table 3.6 is determined as follows:

Lower class boundary = (124+125)/2 = 124.5 g

Upper class boundary = (127+128)/2 = 127.5 g

3.3.2 Width of a Class Interval

The **width of a class interval** is the difference between the lower and upper class boundaries for that class interval. The class width is also called the **class size** or the **class length**.
Thus,

Class width = Upper class boundary – Lower class boundary

Therefore, width of the 4th class interval = 127.5 – 124.5 = 3 grams

Note that if all the class intervals in a grouped frequency distribution have equal widths, then the class width is equal to the difference between two successive lower class limits or two successive upper class limits.
Also, in a grouped frequency distribution, if the class intervals are all the same size or the class marks (class midpoints) are regularly

spaced, then the class interval size or width is equal to the difference between successive class marks (class midpoints), and the class boundaries are midway between the class marks (class midpoints).

To summarize,

- Class interval is bounded by the lower and upper class limits.
- Class interval width is bounded by the lower and upper class boundaries.
- Class mark or class midpoint is the middle of the class interval or class interval width.

Statistical Terms

frequency distribution table. A tabular representation of data by class and the number of observations falling into each class

qualitative frequency distribution table. A frequency distribution table in which the class contains descriptive information

quantitative frequency distribution table. A frequency distribution table in which the class contains numerical values

discrete variable. A variable that can have only certain specified numerical values

continuous variable. A variable that can have any value between two given numerical (measurement) values

grouped frequency distribution. A frequency distribution table in which the class contains ranges of numerical values

class interval. A range of values into which recorded data can be placed

class limits. The lower and upper values of a class interval

lower class limit. The smaller value of a class interval

upper class limit. The larger value of a class interval

class mark or class midpoint. The midpoint of the class interval

lower class boundary. A value slightly smaller than the lower class limit and which does not coincide with any recorded measurement

upper class boundary. A value slightly greater than the upper class limit and which does not coincide with any recorded measurement

class interval width (class size or class length). The difference between the upper and lower class boundaries of a class interval

CHAPTER **4**

Presenting Quantitative Data as Histograms and Frequency Polygons

4.1 Constructing Histograms

A **histogram** is a graphical method used to represent frequency distributions of quantitative data. It bears some resemblance to a vertical bar (column) graph but has numerical values on both the x axis and the y-axis. The histogram is drawn by plotting frequency on the y-axis against class or class interval width on the x-axis. Recall the class may be either single values plotted on the x-axis for ungrouped data, or it may be class interval widths for grouped data. The histogram consists of a set of vertical rectangles with centres at the class mark or class midpoint (for grouped frequency distributions) and widths equal to the class interval widths. The area of each bar is proportional to the respective class frequency. If the classes are all of the same width (for grouped data), then all the bars will be of the same width. Therefore, the heights of the bars are numerically equal to the class frequencies.

4.1.1 Histograms for Discrete Distributions

Recall that a discrete distribution is one where the variable being measured can only have certain definite values. Consider four coins being tossed simultaneously 100 times and after each toss, the number of heads are recorded. The results are summarized in table 4.1.

The information in table 4.1 is interpreted as follows: For each 6 tosses, none of the 4 coins landed with the head up. For each 19 tosses, only 1 coin of the 4 coins landed with the head up. For each 40 tosses, only 2 coins of the 4 coins landed with the head up. For each 31 tosses, 3 coins of the 4 coins landed with the head up. And for each 4 tosses, all 4 of the coins landed with the head up.

Since the number of heads obtained can only be a whole number from zero to four, the data represents a discrete distribution. The information in table 4.1 is represented as a histogram shown in figure 4.1.

Table 4.1 Frequency Distribution of No. of Heads Obtained as a Result of Tossing 4 Coins 100 Times

Number of heads (Class)	Number of tosses (Frequency)
0	6
1	19
2	40
3	31
4	4
Total	100

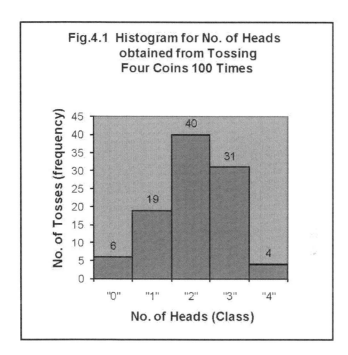

Fig.4.1 Histogram for No. of Heads obtained from Tossing Four Coins 100 Times

Since each bar has a width of one unit, and the height of each bar is equal to the frequency, then the area of each bar is numerically equal to the respective frequency. The sum of the areas of the bars is 6+19+40+31+4 = 100 and is the total area of the histogram bars, which is equal to the total frequency of 100.

4.1.2 Histograms for Grouped Frequency Distributions

Example 4.1

The frequency distribution of the lifetimes of a new type of bulb under development at the Longlife Light Company is shown below in table 4.2.

Table 4.2 Frequency Distribution of the
Lifetime of Bulbs

Lifetime (Hours)	No. of bulbs (Frequency)
400–490	10
491–581	50
582–672	62
673–763	80
764–854	60
855–945	55
946–1036	45
1037–1127	30
1128–1218	8
Total	400

Determine the following:

(a) The class interval of the third class
(b) The upper class limit of the fifth class
(c) The lower class limit of the seventh class
(d) The class boundaries of the eighth class
(e) The class mark of the fourth class
(f) The class interval size
(g) Construct a histogram for the frequency distribution.

Solution 4.1

Table 4.2 shows that 10 bulbs remained lighted continuously for 400 to 490 hours before failing, 50 bulbs remained lighted continuously for 491 to 581 hours before failing, etc.

(a) The class interval of the 3rd class is 582–672.
(b) The upper class limit of the 5th class is 854.
(c) The lower class limit of the 7th class is 946.

(d) The class boundaries of the 8th class are found as follows:

Lower class boundary of 8th class = ½ (1036+1037) = 1036.5
Upper class boundary of 8th class = ½ (1127+1128) = 1127.5

(e) The class mark (or class midpoint) of the 4th class is given by

Class mark of 4th class = ½ (673+763) = 718

(f) The class interval size = (upper class boundary – lower class boundary) for a given class interval.

Consider the 8th class interval.
The class interval size of the 8th class interval = 1127.5–1036.5 = 91

In this case, all class intervals have the same size, i.e., 91 hours.

Take note that in a grouped frequency table, the entries in the class column may be either given as class intervals or class interval widths. If they are class intervals, then each class interval is bounded by lower and upper class limits, in which case the upper class limit of any class interval is not equal to the lower class limit of the next higher class interval. However, if the entries in the class column are class interval widths, then each class interval width is bounded by lower and upper class boundaries, in which case the upper class boundary of any class interval width is equal to the lower class boundary of the next higher class interval width.

When presenting grouped data as a histogram, the width of the vertical bars are equal to the class interval widths and therefore, the sides of the bars are the lower and upper class boundaries for each class interval width. Histograms are drawn based on class boundaries and **not** class limits.

Histograms may also be drawn for grouped distributions using the midpoints of class intervals as the centres of the vertical rectangles, which form the bars.

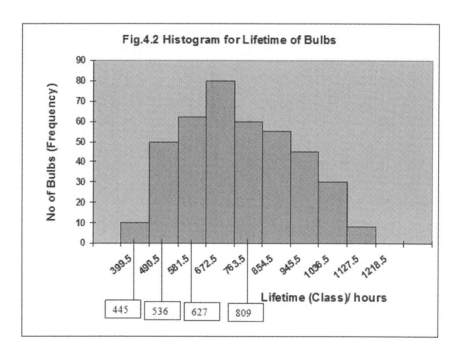

The histogram for the grouped frequency distribution in table 4.2 is shown above in figure 4.2.

Revisiting the frequency table 4.2, the first class interval is 400–490. Therefore, the lower and upper class limits are 400 and 490 respectively, but in the histogram representing table 4.2, the sides of the first bar are not located at 400 and 490 on the horizontal axis but instead at 399.5 and 490.5, which are the lower and upper class boundaries for the first class interval. Similarly, the sides of the fifth bar are not located at 764 and 854, which are the class limits for the fifth class interval, but are located at 763.5 and 854.5, which are the lower and upper class boundaries of the fifth class interval. Thus, for histograms, the sides of the vertical bars are located at the class boundaries for each class interval and not the class limits. As an exercise, the reader should perform the calculations to verify the class boundaries stated in figure 4.2.

Figure 4.2 also shows that the centre of the rectangular bars are located at the class mark (i.e., class midpoint) of each class interval. The class marks for the first, second, third, and fifth bars are shown in the rectangular boxes below the horizontal axis. Section 3.3.1 may be revisited for the method to calculate the class midpoint.

Example 4.2

The following gives the weights of 66 cups of 175 ml ice cream measured to the nearest gram.

106	100	102	102	100	101
101	100	101	98	99	97
107	100	101	102	101	101
101	101	101	98	99	96
102	103	103	103	102	106
100	104	104	104	101	101
101	105	105	101	100	98
108	108	109	101	97	107
101	109	102	101	109	107
101	101	101	102	98	101
97	103	104	101	105	100

(a) Determine the class intervals required to organize the data into 7 groups of equal class intervals.
(b) Determine the class boundaries for each class interval.
(c) Construct a frequency distribution table for the class boundaries.
(d) Draw a histogram to represent the frequency distribution.

Solution 4.2

(a) To determine the class intervals, we need to know the smallest and the largest weights in the data set. Examining the data set, the smallest weight is 96 g, and the largest weight is 109 g. Since the data is to be grouped into 7 class intervals, then the class interval width is given by

Class interval width = Range / No. of class intervals
$$= (\text{largest weight} - \text{smallest weight})/$$
No. of class intervals
$$= (109-96)/7 = 1.8$$
$$= 2.0 \text{ approx.}$$

Starting with the smallest measurement in the data set (96 g) as the lower class limit of the 1st class interval, the 7 class intervals are

96–97
98–99
100–101
102–103
104–105
106–107
108–109

(b) The lower class boundary of the 2nd class interval is (97+98)/2 = 97.5.

The upper class boundary of the 2nd class interval is (99+100)/2 = 99.5.

Therefore, the class interval width = 99.5–97.5 = 2.0.

Since the lower class boundary of the 2nd class interval is equal to the upper class boundary of the 1st class interval, i.e., 97.5, then the lower class boundary of the 1st class interval is 97.5–2.0 = 95.5.

By similar calculation, the class boundaries of each of the class intervals are determined. The class boundaries for the corresponding class intervals are

95.5–97.5
97.5–99.5
99.5–101.5

101.5–103.5
103.5–105.5
105.5–107.5
107.5–109.5

(c) To construct a frequency distribution table, we need to determine the number of observations (measurements) falling within each class. This is done by tallying the measurements in a tally sheet. The result is shown below in table 4.3.

Table 4.3 Frequency Table of Ice Cream Cup Weights

Class intervals	Tally	Frequency
95.5–97.5	////	4
97.5–99.5	### //	6
99.5–101.5	### ### ### ### ### ///	28
101.5–103.5	### ### /	11
103.5–105.5	### //	7
105.5–107.5	###	5
107.5–109.5	###	5
Total		66

(d) The histogram for the frequency distribution of ice cream cup weights is shown in figure 4.3.

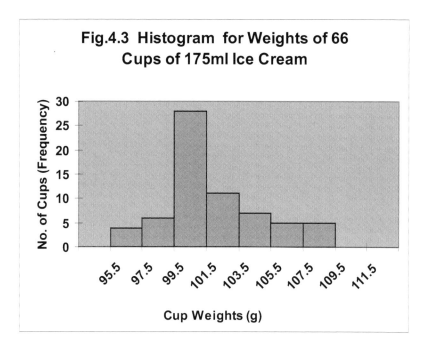

Fig.4.3 Histogram for Weights of 66 Cups of 175ml Ice Cream

When grouped data are presented as a histogram, the width of the vertical bars are equal to the class interval widths, and therefore, the sides of the bars are the lower and upper class boundaries for each class interval width.

If the above problem had simply asked to construct the histogram for the raw data presented, it would not have been necessary to determine the class intervals before ascertaining the class boundaries. The class boundaries could have been found immediately once the class interval width was calculated. To determine the class boundaries directly, take note that the measurements in the data set are to the nearest gram (i.e., integer values). What this means is that a given measurement of say 99 g actually includes all measurements from 98.5 g to 99.5 g. This becomes important in selecting the starting point of the lower boundary of the first class boundary, as it must be chosen so that no measurement in the data set falls on a stated class boundary. In order to make this so, the class boundaries may be stated to one decimal place as no measurement in the data set is to one decimal place and hence no measurement would fall on a class boundary.

As a result, the lower boundary of the first class boundary would be started a little below the smallest weight measurement and rounded off to one decimal place and then incremented by the class interval width, 2.0 g. For example, if 95.5 is selected as the starting point, the class boundaries would be 95.5 g to 97.5 g, 97.5 g to 99.5 g, 99.5 g to 101.5 g, etc. Other suitable starting points could have been 95.1 g, 95.2 g, 95.7 g, or any value to one decimal place as there are no measurements recorded to one decimal place, and all class boundaries would be rounded off to one decimal place.

4.2 Frequency Polygons

A **frequency polygon** is another way of representing a frequency distribution. The frequency polygon is obtained by connecting the midpoints of the tops of the rectangles in a histogram by straight lines. If, however, the midpoints of the tops of the rectangles in a histogram are joined by a smooth curve, the polygon formed is called a **smoothed frequency polygon**.

Consider the histogram in figure 4.3, which is drawn below as figure 4.4, with the frequency polygon superimposed on it. It is customary to add the extensions AB and CD to the next lower and higher class midpoints at the ends of the diagram, which are each assigned a corresponding class frequency of zero. When this is done, the area under the frequency polygon (i.e., the area between the polygon curve and the x-axis) equals the area of the rectangles of the histogram.

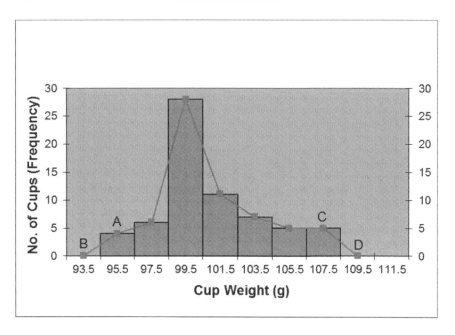

Fig. 4.4 Histogram for Weights of 66 Cups of 175ml Ice Cream

Alternatively, the frequency polygon may be drawn by plotting the midpoint of each class interval on the horizontal axis against the frequency of that class on the vertical axis. The points are then connected with a series of line segments. As before, the polygon is completed by drawing a line segment from the point of intersection of the class mark and the class frequency of the largest class to the point of intersection of the class mark of the next higher class and the horizontal axis and another line segment from the point of intersection of the class mark and the class frequency of the smallest class to the point of intersection of the class mark of the preceding smaller class and the horizontal axis.

4.2.1 Some Types of Frequency Distributions

Histograms of frequency distributions may have one of three general shapes. They may be symmetrical, skewed to the right, or skewed to the left (see fig. 4.5). When the shape of the histogram is symmetrical or close to symmetrical as in figure 4.5a, the frequency distribution is described as a **normal distribution**.

Since frequency distributions may also be represented graphically as smoothed frequency polygons, figure 4.5a also shows the smoothed frequency polygon superimposed on the histogram. Figure 4.6 shows symmetrical and skewed frequency distributions as smoothed frequency polygons.

Figure 4.6a shows the smoothed frequency polygon that is symmetrical. This type of curve is referred to as a **bell-shaped curve** or a **normal curve**. *Frequency distributions that form a bell-shaped curve are called normal distributions.* Many real-life situations approximate this normal

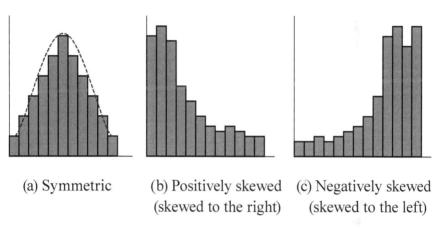

| (a) Symmetric | (b) Positively skewed (skewed to the right) | (c) Negatively skewed (skewed to the left) |

**Fig. 4.5 Skewness of Frequency Distributions
Shown as Histograms**

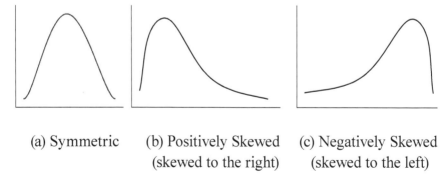

| (a) Symmetric | (b) Positively Skewed (skewed to the right) | (c) Negatively Skewed (skewed to the left) |

**Fig. 4.6 Skewness of Frequency Distributions
Shown as Smoothed Frequency Polygons.**

distribution, for example, the heights of people within some age group or the weight of a product from a production line process. Consider a group of people. The majority of persons would be within some average height range, and there would be a few persons taller than this average height range and also a few persons shorter than this height range. In the case of a stable production process, consider the production of 1-litre containers of vanilla ice cream, which are to have a target weight of say 1300 grams. If the weights of 100 units are determined, some will have weights greater than the target weight of 1300 grams, while some will have weights lighter than that, but the majority would be close to the target weight.

Frequency distributions that show lack of symmetry may be skewed to the right (fig. 4.6b) or skewed to the left (fig. 4.6c). A frequency distribution that is skewed to the right or positively skewed has a long "tail" extending to the right. A frequency distribution that is skewed to the left or negatively skewed has a long tail extending to the left.

Statistical Terms

histogram. A graphical method of presenting a frequency distribution table for variable data. It consists of vertical bars having a width equal to the class interval width and height equal to the corresponding class frequency. The bars share common sides with no space between them.

frequency polygon. A graphical method of presenting a frequency distribution for variable data. It is constructed by plotting the midpoints of the class intervals against the corresponding class frequency and joining these points by straight line segments. To complete (or close) the polygon, two more class midpoints are added, one preceding the smallest given class midpoint and one proceeding the largest given class midpoint. Both of these are assigned a zero (0) frequency value, and straight line segments are drawn to the horizontal axis of these added class marks.

smoothed frequency polygon. A frequency polygon drawn as a smooth curve instead of straight line segments

normal distribution. A frequency distribution that, when plotted, yields a bell-shaped curve.

bell-shaped curve or normal curve. A smoothed frequency polygon that is symmetrical

skewed frequency distribution. A frequency distribution that, when plotted, shows lack of symmetry by having a long tail extending to the right or to the left.

CHAPTER 5

Cumulative Frequency Distributions and Polygons

5.1 Cumulative Frequency Distributions

In many instances, it is necessary to know how many measurement values in a data set are less than or equal to some *particular* measurement value. This number of measurement values is called the **cumulative frequency** of the *particular* measurement value. The cumulative frequency is obtained by adding up the frequencies of all measurement values less than the *particular* measurement value plus the frequency of the *particular* measurement value.

Example 5.1

The table shows the heights of 60 people. Determine the cumulative frequency for each height, and draw the cumulative frequency polygon.

Table 5.1 Frequency Distribution of Heights of 60 People

Height (cm)	Frequency
165	2
166	7
167	11
168	17
169	12
170	8
171	3

Notice that table 5.1 is a *quantitative* frequency distribution table. It is described as a quantitative table because the variable being measured and which forms the class of the frequency distribution table is height, which is measured and recorded as a numerical value.

Solution 5.1

The table below shows the cumulative frequencies for the various heights.

Table 5.2 Cumulative Frequency Distribution
of Heights of 60 People

Height (cm)	Frequency	Cumulative frequency
165	2	2
166	7	9
167	11	20
168	17	37
169	12	49
170	8	57
171	3	60

Note that each entry in the cumulative frequency column is obtained by adding the preceding entries in the frequency column. Thus, 9 = 2+7, 20 = 2+7+11, 37 = 2+7+11+17, etc.

Fig. 5.1 Cumulative Frequency Polygon for Heights of 60 Persons

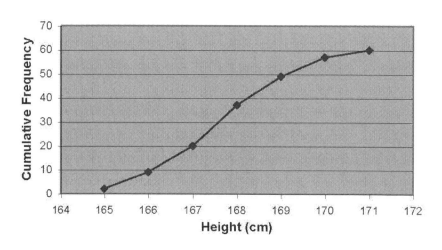

From the table, it can be seen that for a height of 168 cm, the cumulative frequency is 37. This tells us that there are 37 height measurements that are smaller than or equal to 168 cm, or alternatively, there are 37 persons out of the 60 persons who are 168 cm tall and shorter. The cumulative frequency polygon is obtained by plotting the cumulative frequency for each height measurement against the respective height measurement and joining the points by straight line segments. The plot is shown in figure 5.1.

Example 5.2

Construct a cumulative frequency distribution, a percentage cumulative distribution, and draw both the cumulative frequency polygon and percentage cumulative frequency curve for the grouped frequency distribution shown below in table 5.3.

Table 5.3 Frequency Table for 175 ml Ice Cream
Cup Weights

Cup weights (g) (Class interval)	Frequency
95.5–97.5	4
97.5–99.5	6
99.5–101.5	28
101.5–103.5	11
103.5–105.5	7
105.5–107.5	5
107.5–109.5	5

Solution 5.2

You might notice that table 5.3 is table 4.3 from the previous chapter, which summarises the weights of 66 cups of 175 ml ice cream. The cumulative frequency of the various classes is shown below.

Table 5.4 Cumulative Frequency Table for
175 ml Ice Cream Cup Weights

Cup weights (g) (Class interval)	Frequency	Cumulative frequency
95.5–97.5	4	4
97.5–99.5	6	4+6 = 10
99.5–101.5	28	10+28= 38
101.5–103.5	11	38+11= 49
103.5–105.5	7	49+7 = 56
105.5–107.5	5	56+5 = 61
107.5–109.5	5	61+5 = 66

It must be realized that although it is convenient to list the cumulative frequency corresponding to the associated class interval, this is not strictly correct. To explain, the class interval 101.5–103.5

has a cumulative frequency of 49. Because we are talking about cumulative frequency, this really means that there are 49 measurements (or 49 cups) whose weight is less than 103.5 g. That is, there are 49 measurements that lie in the class interval 101.5–103.5 and all lower class intervals. Notice that in the sentence before, I did not say less than or equal to 103.5 g because in this case, we are dealing with a grouped distribution with class boundaries, and remember that class boundaries are chosen so that no actual measurement falls on a class boundary, so there is no cup weighing 103.5 g. The table below shows the correct way to describe the class intervals when dealing with cumulative frequencies.

Table 5.5 Correct Way of Describing a
Cumulative Frequency Table

Cup weights (g) (Class interval)	Frequency	Cumulative Frequency	%Cumulative Frequency
Less than 97.5	4	4	6.1
Less than 99.5	6	4+6 = 10	15.2
Less than 101.5	28	10+28= 38	57.6
Less than 103.5	11	38+11= 49	74.2
Less than 105.5	7	49+7 = 56	84.4
Less than 107.5	5	56+5 = 61	92.4
Less than 109.5	5	61+5 = 66	100.0

Note that in obtaining the cumulative frequency distribution, the upper boundary limit of each class was used. It is perfectly acceptable to also use the lower class boundary of each class if desired. From table 5.5, it can be seen, for example, that 38 cups of ice cream weigh less than 101.5 grams.

The percentage cumulative distribution is also shown in column 4 in table 5.5 and is obtained by dividing the respective cumulative frequency value in the previous column by the total frequency (66) and expressing it as a percentage. For example, the percent cumulative frequency of 74.2% = 49/66 x 100%.

A cumulative frequency distribution may be represented graphically by plotting the cumulative frequency against the upper class boundary and joining the points by straight line segments or a smooth curve to form a **cumulative frequency polygon** or a **cumulative frequency curve** respectively. The upper class boundaries are plotted on the horizontal axis. The cumulative frequency curve is also called an ogive after the architectural term for this type of curve. The cumulative frequency polygon and cumulative frequency curve are shown in figure 5.1 and figure 5.2.

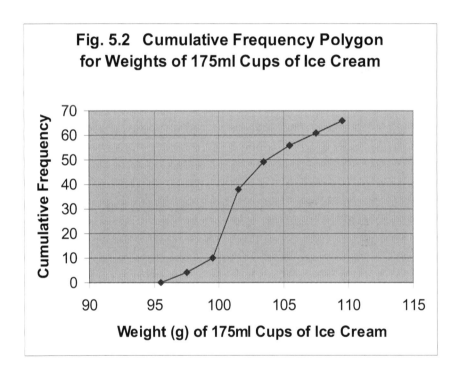

From figure 5.2, it can be seen that the cumulative frequency corresponding to a cup weight of 100 g is about 17. Thus, 17 cups of ice cream have a weight of 100 g or less. Also, the cumulative frequency corresponding to a weight of 105 g is about 55. Therefore, 55 cups of ice cream have a weight of 105 g or less. Conversely, since 66 cups make up the sample data set, it can be said that 66-55 = 11 cups of ice cream have a weight greater than 105 g.

Since the cumulative frequencies corresponding to cup weights of 100 g and 105 g are 17 and 55 respectively, the number of cups having a weight between 100 g and 105 g inclusive is 55−17= 38.

Figure 5.3 below shows the percent cumulative frequency curve, which is very similar to the cumulative frequency polygon with the exception that the points are joined by a smooth curve instead of straight line segments as in figure 5.2, and the vertical scale is expressed as percent cumulative frequency.

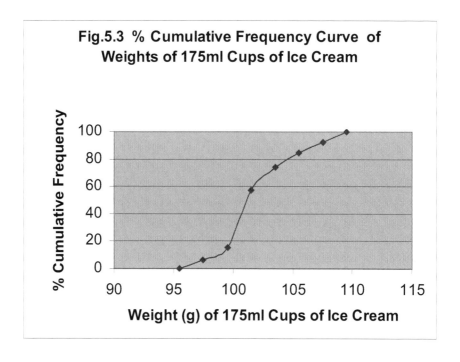

Fig.5.3 % Cumulative Frequency Curve of Weights of 175ml Cups of Ice Cream

Statistical Terms

cumulative frequency. The number of measurement values that are less than or equal to some particular measurement value. It is evaluated for any given measurement value by summing all frequencies for all measurements that are smaller than or equal to that given measurement value.

cumulative frequency polygon. A graph obtained by plotting the cumulative frequency against the respective measurement value and joining these points by straight line segments

cumulative frequency curve. A graph obtained by plotting the cumulative frequency against the respective measurement value and joining these points by a smooth curve

percent cumulative frequency polygon (curve). A graph obtained by plotting the percent cumulative frequency against the respective measurement value and joining these points by straight line segments or a smooth curve

CHAPTER 6

Relative Frequency and Cumulative Relative Frequency Distributions

6.1 Relative Frequency

Consider a sample of size n, having a particular item in the sample with a measurement x that occurs f times, i.e., the frequency of the measurement x is f. The quotient f/n is called the **relative frequency** of that measurement x. If now, for that x measurement, all frequencies corresponding to values smaller than or equal to x are summed, the result is the cumulative frequency corresponding to that x value. When this cumulative frequency is now divided by the sample size n, this yields the **relative cumulative frequency** for that x value.

Alternatively, if for that same x measurement, all relative frequencies corresponding to values smaller than or equal to x are summed, the result is the **cumulative relative frequency** for that same x value. Note that the relative cumulative frequency is equal to the cumulative relative frequency.

Example 6.1

Five 25-cent coins were tossed 125 times, and after each toss, the number of heads obtained were recorded. The number of tosses resulting in 0, 1, 2, 3, 4, and 5 heads are shown in table 6.1.

Table. 6.1 Frequency Distribution of No. of Heads
Obtained in 125 Tosses of 5 Coins

No. of heads	No. of tosses (frequency)
0	5
1	18
2	42
3	38
4	19
5	3
Total	125

(i) Construct a table of (a) relative frequency, (b) cumulative frequency, and (c) relative cumulative frequency.

(ii) Draw (a) a line graph, (b) a frequency histogram, (c) a frequency polygon, (d) a relative frequency histogram, (e) a relative frequency polygon, (f) cumulative frequency distribution, and (g) a relative cumulative frequency distribution.

Solution 6.1

Note that table 6.1 is a data set for discrete data as the variable *no. of heads* is a discrete variable as it can only take on integer values from 0 to 5. The sample size *n* is 125, which is equal to the sum of the frequencies. The sum of the frequencies is denoted by Σf.

The symbol Σ is the Greek capital letter called sigma and is used in mathematics to denote summation.

(i) The solution is summarised in table 6.2 below.

Table 6.2 Required Solution to Part (i)

No. of heads (class)	No. of tosses (frequency) (f)	Relative frequency (f/Σf) or (f/n)	Relative frequency %	Cumulative frequency	Relative cumulative frequency	Relative cumulative frequency %
0	5	0.040	4.00	5	0.040	4.00
1	18	0.144	14.4	23	0.184	18.4
2	42	0.336	33.6	65	0.520	52.0
3	38	0.304	30.4	103	0.824	82.4
4	19	0.152	15.2	122	0.976	97.6
5	3	0.024	2.40	125	0.100	100.0
Total	125	1.00	100			

The relative frequency is obtained by dividing the frequency for each class by the total frequency of all the classes. [Note: the sum of the frequencies for all the classes is equal to the sample size (n)]. That is,

$$f = \text{frequency for any class}$$
$$\Sigma f = \text{sum of all frequencies} = n = \text{sample size}$$

Thus the relative frequency for obtaining two heads is $42/125 = 0.336$. The relative frequency is interpreted as the proportion or fraction of the total number of observations falling into each class. It may be expressed as a decimal fraction as in column 3 in table 6.2 or as a percentage as in column 4 of table 6.2. For example, it can be said that 33.6% of the 125 tosses resulted in 2 coins of the 5 coins landing with the head up.

The cumulative frequency, you would recall, is the number of measurement values that are less than or equal to some particular measurement value. It is obtained by summing all frequencies with a measurement less than or equal to the measurement in question. The cumulative frequency for the various classes are shown in column

5 of table 6.2. The cumulative frequency of 103 means that for 103 tosses, 3 or fewer coins landed with head uppermost.

The relative cumulative frequency is the proportion or fraction of items with measurement values less than or equal to some particular measurement or upper boundary of a given class interval for grouped data. It is calculated by dividing the cumulative frequency for each class by the sample size n, which is also the sum of all frequencies, Σf. Relative cumulative frequency may be expressed as a decimal fraction as in column 6 in table 6.2 or as a percentage as in column 7 in table 6.2. For example, the relative cumulative frequency of 82.4% means that 82.4% of the tosses resulted in 3 or fewer coins landing with the head up.

It is left as an exercise for the reader to verify that the cumulative relative frequency is equal to the relative cumulative frequency.

(ii) (a)   The line graph is shown in figure 6.1. Since the data for this problem is discrete, the line graph seems to be the most natural graph to represent the data. The graph is set up similar to a bar graph, but the bars have zero width.

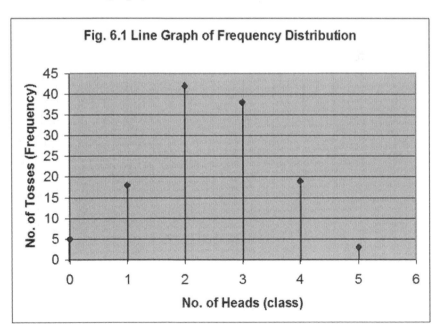

Fig. 6.1 Line Graph of Frequency Distribution

(ii) (b)   Figure 6.2 shows the frequency histogram for the data.
The histogram representation of the data treats the data as
continuous. Note that the area of the histogram is the total
frequency—125.

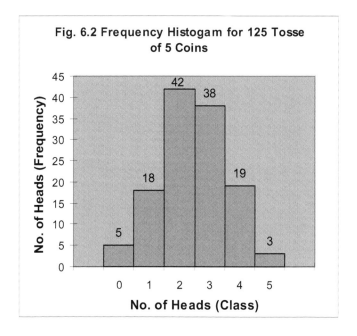

Fig. 6.2 Frequency Histogam for 125 Tosse
of 5 Coins

(ii) (c) Figure 6.3 shows the frequency histogram with the
frequency polygon superimposed on it. Recall that the area

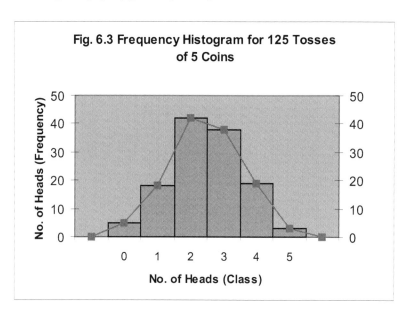

Fig. 6.3 Frequency Histogram for 125 Tosses
of 5 Coins

under the polygon must be equal to the area of the histogram. To make this so, straight line segments must be added from the top centre of each end bar to the x-axis. The horizontal projection of the added line segments must be equal to the bar width.

Figure 6.4 shows the frequency polygon by itself for the data. The frequency polygon also treats the data as if it were continuous.

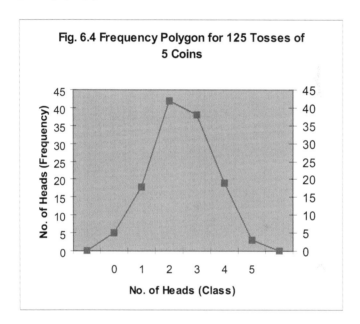

(ii) (d and e) Figure 6.5 shows the relative frequency histogram for the 125 tosses of 5 coins. The relative frequency is expressed as a decimal fraction. The figure also shows the relative frequency polygon. Fig. 6.6 shows the relative frequency graph expressed as a percentage.

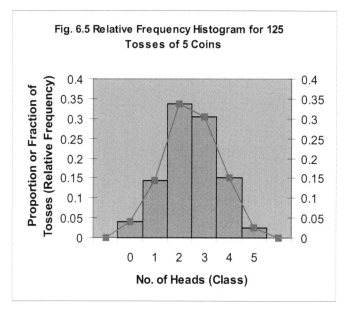

Fig. 6.5 Relative Frequency Histogram for 125 Tosses of 5 Coins

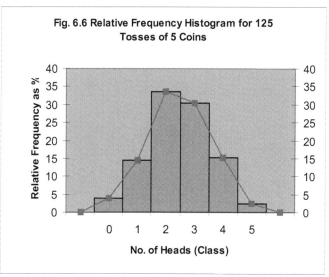

Fig. 6.6 Relative Frequency Histogram for 125 Tosses of 5 Coins

Notice that the graphs for relative frequency, whether expressed as a decimal fraction or a percentage, are similar to each other but only have different scales for the y-axis.

(ii) (f) Figure 6.7 below shows the cumulative frequency graph
for the 125 tosses of 5 coins.

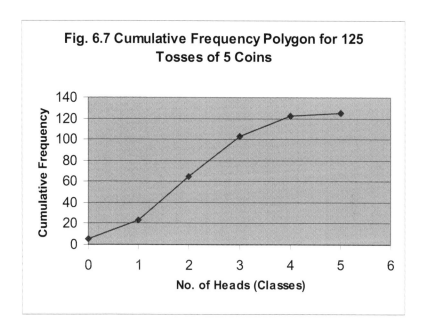

(ii) (g) Figure 6.8 shows the relative cumulative frequency curve,
also called the cumulative relative frequency curve, expressed
as a percentage.

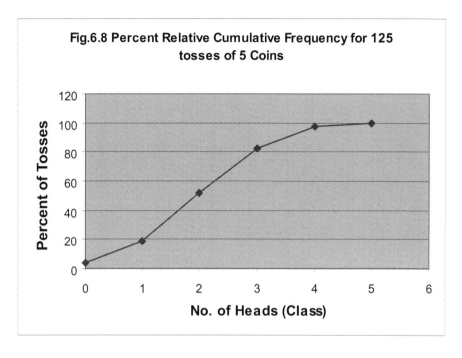

Fig.6.8 Percent Relative Cumulative Frequency for 125 tosses of 5 Coins

Example 6.2

The frequency distribution of the lengths of a sample of rods measured to the nearest millimetre (mm) is given in table 6.3 below.

Table 6.3 Frequency Distribution of Rod Lengths

Length (mm)	Frequency
200 – 204	5
205 – 209	9
210 – 214	13
215 – 219	18
220 – 224	44
225 – 229	28
230 – 233	26
234 – 239	7

(a) Determine the number of rods in the sample.

(b) Construct the corresponding relative frequency, cumulative frequency, and relative cumulative frequency tables.
(c) Draw (i) the frequency histogram;
  (ii) the frequency polygon;
  (iii) the percent relative frequency histogram;
  (iv) the cumulative frequency polygon; and
  (v) the percent relative cumulative frequency polygon.

Solution 6.2

(a) The number of rods in the sample is equal to the sample size n, which is in turn equal to the sum of all the frequencies Σf.

Therefore, $n = \Sigma f = 5+9+13+18+44+28+26+7 = 150$ rods.

(b) (i) The relative frequency table is shown in table 6.4.

The relative frequency is calculated for each class interval by dividing the class frequency by the sample size.

Consider the class interval "210 – 214",
Class frequency= 13, and sample size = 150,
Therefore, relative frequency = 13/150 = 0.087 or 8.7%.

Table 6.4 Relative Frequency Distribution of Lengths of 150 Rods

Length (mm) (Class)	Frequency	Relative Frequency	% Relative Frequency
200 – 204	5	0.033	3.3
205 – 209	9	0.060	6.0
210 – 214	13	0.087	8.7
215 – 219	18	0.120	12.0
220 – 224	44	0.293	29.3
225 – 229	28	0.187	18.7
230 – 233	26	0.173	17.3
235 – 239	7	0.047	4.7

(b) (ii) Before the cumulative frequency can be determined, the upper class boundary for each class interval must be found. Thus,

upper class boundary for class interval 200 – 204 = ½(204+205) = 204.5
upper class boundary for class interval 205 – 209 = ½(209+210) = 209.5
upper class boundary for class interval 210 – 214 = ½(214+215) = 214.5, etc.

Having determined the upper class boundary for each class interval, the cumulative frequency for each class boundary of a given class interval is found by summing the frequencies for all class intervals less than or equal to that given class interval.

The cumulative frequency table is shown in table 6.5.

The relative cumulative frequency is then found for each class by dividing the respective cumulative frequency by the sample size and multiplying by 100 to express it as a percentage.

Table 6.5 Cumulative Frequency and Percent Relative Cumulative Frequency of Lengths of 150 Rods

Length (mm) (Class)	Cumulative frequency	Relative cumulative frequency %
Less than or equal to 204.5	5	3.3
Less than or equal to 209.5	14	9.3
Less than or equal to 214.5	27	18.0
Less than or equal to 219.5	45	30.0
Less than or equal to 224.5	89	59.3
Less than or equal to 229.5	117	78.0
Less than or equal to 233.5	143	95.3
Less than or equal to 239.5	150	100.0

(c) (i) Figure 6.9 shows the frequency histogram with the frequency polygon superimposed on it. Recall that the area under the polygon must be equal to the area of the histogram. To make this so, straight line segments must be added from the top centre of the end bars to the x-axis. The horizontal projection of the added line segments must be equal to the bar width.

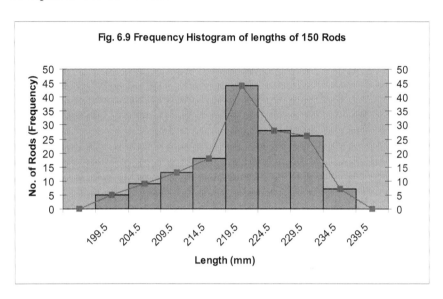

Recall also that for grouped data, the sides of each bar for a given class interval lie at the lower and upper class boundaries for that class interval, and the centre of the bars lie at the class midpoint or class mark of the class intervals.

(d) (ii) The frequency polygon is shown by itself in figure 6.10. The frequency polygon can be drawn directly by plotting the class frequency against the class midpoint (class mark).

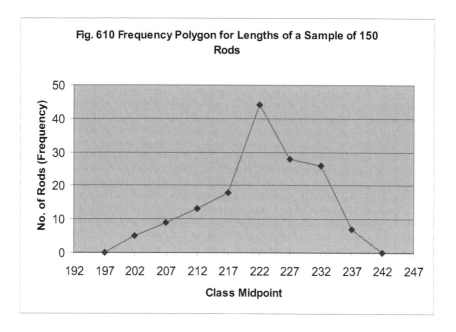

(c) (iii) Figure 6.11 shows the relative frequency histogram with the relative frequency polygon superimposed on it. Note again that the

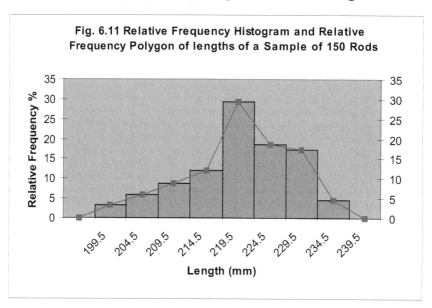

graph is similar to the frequency histogram with the scale on the vertical axis adjusted for the relative frequency values expressed as a percentage.

(c) (iv) The cumulative frequency curve is shown in figure 6.12. The curve is a plot of cumulative frequency against the upper class boundary for each class interval.

(c) (v) The percent relative cumulative frequency curve is shown in figure 6.13.

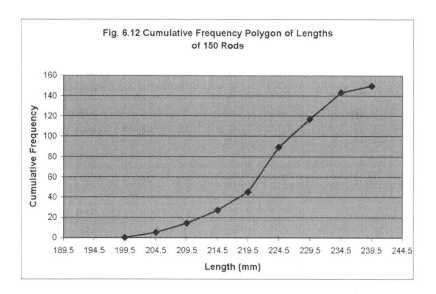

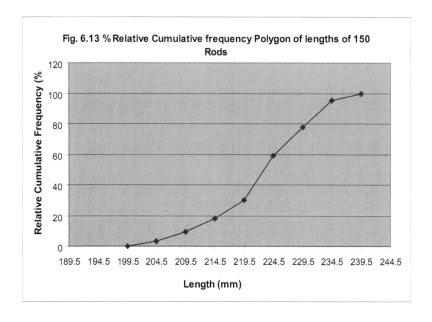

Statistical Terms

relative frequency. The proportion or fraction of the total number of observations (measurements) falling in each class

percent relative frequency. The percentage of the total number of observations (measurements) falling in each class

relative cumulative frequency. The proportion or fraction of the total number of observations with a value less than or equal to a particular observation or measurement value

percent relative cumulative frequency. The percentage of the total number of observations (measurements) with a value less than or equal to a particular observation or measurement value

CHAPTER **7**

Describing Data Using Numerical Descriptive Measures

7.1 Types of Numerical Descriptive Measures

In the previous chapters, we presented graphical statistical methods to help organise, summarise, and represent data sets so that we could make sense of and interpret the data collected. These graphical or pictorial representations of a data set gave a good visual overview of the relationships and patterns of the distribution at a glance. In this chapter, we present another statistical method to help analyse and interpret data sets. With this method, a few *special numbers* about the data set are calculated, which are then used to describe or summarise the characteristics of the data set. These *special numbers* are referred to as **numerical descriptive measures** or **descriptive statistics** and are used to form a mental picture of the data set's distribution. These numerical descriptive measures include the following:

1) **Measures of central tendency** — This is a number that describes the centre of the data set (distribution) and is located near the spot where most of the data tend to cluster.

2) **Measures of dispersion** — This is a number that measures the amount of spread the measurements in a data set have.
3) **Measures of relative standing** — This is used to determine where a particular measurement in a data set lies relative to all other measurements in the data set.
4) **Measures of association** — This is used to determine if there is a relationship or correlation between two different variables in a data set.

Consider a data set of test scores marked out of 100 marks of all students who took a physics examination. The teacher of the subject may be interested in knowing the score where most of the marks cluster and whether this clustering was towards the lower end of the marking scale, somewhere in the middle, or toward the high end of the marking scale. This characteristic about the data set, which locates where the majority of the marks are clustered and hence defines the centre of the data distribution, is summarised by the numerical descriptive measure called **measures of central tendency.**

The teacher may also be interested in knowing the amount of spread in the test scores from the lowest mark to the highest mark achieved. This characteristic of the data set, which measures the range of variation among the test scores, is summarised by the numerical descriptive measure called **measures of variation** or **measures of dispersion**.

The teacher may also be interested in determining where a particular score lies in relation to other scores in order to be able to rank a student, e.g., by saying this particular student ranked in the top 5% or the bottom third, etc. This characteristic of the data set, which determines where a particular score lies relative to the other scores is described by the numerical descriptive measure called **measures of relative standing**.

The teacher may further be interested in examining whether the test scores of the students may be related with other performance measures of the student, such as laboratory work. The question being, Did the students who did well in the written paper also do well in the laboratory coursework? The degree of association between two such measures is described by the numerical descriptive measure called **measures of association**.

We will now take a closer look at these numerical descriptive measures, beginning with measures of central tendency.

7.2 Measures of Central Tendency

Measures of central tendency attempt to locate the centre of a data set. Here, we are seeking to find a number that is at the centre or located close to the middle of the data set or frequency distribution of data. Three measures of central tendency are the following:

(i) mean
(ii) median
(iii) mode

7.2.1 The Mean

The mean is simply the arithmetic average of all the values in a data set. This average is equal to the sum of the values in the sample divided by the number of values in the sample. If, in a sample, there are n observations (measurements) of a variable denoted by $x_1, x_2, x_3, ... x_n$ then the arithmetic average or sample mean of the data set is denoted by $\bar{x}$ (read x bar) and is given by:

$$\bar{x} = \frac{\text{Sum of the } x \text{ values}}{\text{No. of observations}} = \frac{\Sigma x}{n}$$

where, Σx denotes the sum of all measurements. As mentioned before, the symbol Σ is the Greek capital letter called sigma and is used in mathematics to denote summation.

Example 7.1

Calculate the mean of the following measurements: 11, 16, 13, 21, 15, 17, and 13.

Solution 7.1

The number of measurements is $7 = n$, the sample size.

Applying the equation for mean,

$$\bar{x} = \frac{\Sigma x}{n} = \frac{11+16+13+21+15+17+13}{7} = \frac{106}{7} = 15.1$$

Example 7.2

Calculate the mean of the following frequency distribution.

Measurement (x)	14	12	9	8	7
Frequency (f)	2	5	7	4	1

Solution 7.2

In this problem, the data consist of repeated measurements. The measurement 14 occurs twice, the measurement 12 occurs 5 times, the measurement 9 occurs 7 times, the measurement 8 occurs 4 times, and the measurement 7 occurs once. Thus, associated with each measurement x is a frequency f of its occurrence.

The total number of measurements in the frequency table is equal to the sum of the frequencies, i.e.,

Sum of frequencies, $\Sigma f = 2+5+7+4+1$
=19, which is the sample size n.

The mean of the distribution is given by

$$\bar{x} = \frac{\sum x f}{\sum f}$$

$$= \frac{(14)(2)+(12)(5)+(9)(7)+(8)(4)+(7)(1)}{2+5+7+4+1}$$

$$= \frac{28+60+63+32+7}{19} = \frac{190}{19} = 10$$

$$\Rightarrow \bar{x} = 10$$

Example 7.3

Calculate the mean weight of 26 male students shown in the frequency distribution below.

Weight (kg)	Frequency
51 – 55	2
56 – 60	3
61 – 64	4
65 – 69	7
70 – 74	5
75 – 79	3
80 – 84	2
Total	26

Solution 7.3

This problem presents the data as **grouped data** in a frequency distribution table. Note in this case that without access to the raw data, the exact value of the mean cannot be calculated. Take a look at the class intervals in the table. For any of the class intervals, we only know how many weights fall into that interval. We do not know the exact values of the weights. As a result, we cannot add up the x values (i.e., all the weights) because we do not know them. So in order to calculate the mean of a **grouped** frequency distribution, it is assumed that the observations (measurements) in a given class interval are distributed uniformly over the interval and have a mean value equal to the class midpoint (or class mark) denoted by x_m. Thus, the mean of the distribution is calculated using the following formula:

$$\bar{x} = \frac{\sum x_m f}{n}$$

where f is the frequency for the given class interval (i.e. the class frequency).

The class midpoints, as you will remember, is the average of the lower and upper class limits. The table below shows the original table with the class midpoints included.

Weight (kg) (Class)	Class midpoint (x_m)	Frequency (f)	$x_m f$
51 – 55	53	2	106
56 – 60	58	3	174
61 – 64	63	4	252
65 – 69	67	7	469
70 – 74	72	5	360
75 – 79	77	3	231
80 – 84	82	2	164
Total		26	1756

The mean is $\bar{x} = \dfrac{1756}{26} = 67.5$ kg

This calculated mean would be a little different from the mean that would have been calculated had the actual raw data been used.

7.2.2 The Median

The median is a number that divides the data set into two halves, such that half of the observations (measurements) are larger than the median and half are less than the median. For a data set containing an odd number of measurements, the median is the middle measurement obtained after arranging the measurements in either ascending or descending order and is located at the $\left(\dfrac{n+1}{2}\right)$ position, where n is the sample size of the data set or the number of observations (measurements) making up the data set. If the data set contains an even number of measurements, the median is the arithmetic mean of the two middle measurements which are located at the $\left(\dfrac{n}{2}\right)$ and $\left(\dfrac{n}{2}+1\right)$ positions after the measurements are arranged in either ascending or descending order.

Example 7.4

Calculate the median of the following measurements: 11, 16, 13, 21, 15, 17, and 19.

Solution 7.4

The number of measurements is $7 = n$, the sample size.

Arranging the measurements in ascending order gives

$$11, 13, 15, 16, 17, 19, 21$$

Since there is an odd number of measurements, there is only one middle value, which is 16, the median, and is located at the $\left(\dfrac{n+1}{2}\right) = \left(\dfrac{7+1}{2}\right) = 4$, i.e., the 4th position. Notice that there are three measurements greater than 16 and three measurements smaller than 16. Recall that the median divides the data set into two equal halves.

Example 7.5

Calculate the median of the following measurements: 11, 16, 13, 21, 15, 17, 19, and 23.

Solution 7.5

The number of measurements is $8 = n$, the sample size.

Arranging the measurements in ascending order gives

$$11, 13, 15, 16, 17, 19, 21, 23$$

Since there is an even number of measurements, the median is given by the arithmetic mean of the two middle numbers, 16 and 17. They are the two middle numbers located at the $\left(\dfrac{n}{2}\right)$ and $\left(\dfrac{n}{2}+1\right)$ positions, i.e., $\left(\dfrac{8}{2}\right) = 4$ and $\left(\dfrac{8}{2}+1\right) = 5$, the 4th and 5th positions in the data set after being arranged in ascending order.

$$\text{The median} = \left(\frac{16+17}{2}\right) = 16.5.$$

Example 7.6

Calculate the median of the following frequency distribution.

Measurement (x)	14	12	9	8	7
Frequency (f)	2	5	7	4	1

Solution 7.6

In this problem, the data consist of repeated measurements. That is, the measurement 14 occurs twice, the measurement 12 occurs 5 times, the measurement 9 occurs 7 times, the measurement 8 occurs 4 times, and the measurement 7 occurs once. Thus, associated with each measurement x is a frequency f of occurrence.

The total number of measurements in the frequency distribution table is equal to the sum of the frequencies, $\sum f = 2+5+7+4+1 = 19$. Thus, the sample size $n = 19$.

Since there is an odd number of measurements, there is only one middle value, which is located at the $\left(\dfrac{n+1}{2}\right) = \left(\dfrac{19+1}{2}\right) = 10$, i.e., the 10th position.

Notice in this case that the frequency distribution table has the measurement values arranged in descending order from left to right. The sum of the first two class frequencies is $2+5 = 7$. To get to the tenth position requires three more measurements, all of which come from the third class, which has a frequency of 7. The 7 measurements in this third class are all 9, and hence, a measurement of 9 occupies the 10th position. Thus, the required median is 9.

If the measurement values in the frequency table were in random order, then before calculating the median, the measurements would have had to be written in either ascending or descending order.

7.2.2.1 Grouped Data

When data is presented in a grouped format, the values of the individual measurements are lost. The first step in finding the median for **grouped data** is to locate the class interval in which the middle measurement value lies. This class interval is referred to as the **median class.** Note that the measurement intervals must be arranged either in ascending or descending order before calculating the median. After the median class has been determined, the median value is found by interpolation using the following formula

$$\text{Median} = L + \left(\frac{n_1}{f_m}\right) W$$

where   L is the lower class boundary of the median class,

n_1 is the number of measurements in the median class that are below the median and inclusive of the median

f_m is the frequency of the median class

W is the class width of the median class

Example 7.7

Calculate the median weight of 26 male students shown in the frequency distribution below.

Weight (kg)	Frequency
51 – 55	2
56 – 60	3
61 – 64	4
65 – 69	7
70 – 74	5
75 – 79	3
80 – 84	2
Total	26

Solution 7.7

This problem presents the data as a frequency distribution of grouped data. Note in this case that without access to the raw data, the exact value of the median cannot be calculated. Take a look at the class intervals in the table. For any of the class intervals, we only know how many weights fall into that interval. We do not know the exact weight values.

Apply the formula

$$\text{Median} = L + \left(\frac{n_1}{f_m}\right)W$$

where   L is the lower class boundary of the median class,

n_1 is the number of measurements in the median class up to and including the median

f_m is the frequency of the median class

W is the class width of the median class

Since the sample size is $n = 26$, meaning that 26 measurements comprise the sample, then the median or centre measurement occurs at the $\left(\frac{n}{2}\right) = \left(\frac{26}{2}\right) = 13$, the 13th position.

The sum of the first three class frequencies is $2+3+4 = 9$. To get to the thirteenth measurement requires *four* more measurements, all of which come from the fourth class interval that has a frequency of 7. Thus, the median lies in the fourth class interval, which is therefore the median class.

Now, the fourth class interval 65–69 actually corresponds to weights ranging from 64.5 to 69.5 kg, which are the lower and upper class boundaries respectively for that class interval.Thus,

$$L = 64.5,$$
$$n_1 = 4,$$
$$f_m = 7,$$
$$W = 69.5 - 64.5 = 5$$

Substituting in the equation Median $= L + \left(\dfrac{n_1}{f_m}\right) W$ gives

$$\text{Median} = 64.5 + \left(\frac{4}{7}\right)(5) = 67.4 \text{ kg.}$$

Alternatively, n_1 may be found from the following equation:

$$n_1 = \frac{n}{2} - \left(\sum f\right)_1$$

where

$n=26$, the sample size

and $\left(\sum f\right)_1 =$ sum of all class frequencies
below the median class $= 2+3+4 = 9$

Therefore,

$$n_1 = \frac{26}{2} - 9 = 4.$$

7.2.2.2 Finding the Median from a Histogram

Given a histogram, the median is the value of x (the abscissa) such that a vertical line through x divides the histogram into two parts each of equal area.

Example 7.8

Find the median length of a sample of 150 rods using the frequency histogram shown below. This histogram is from example 6.2 and is reproduced here as figure 7.1.

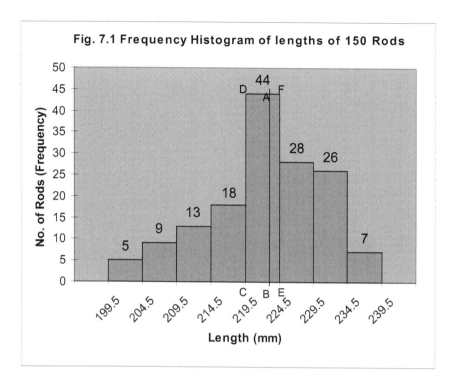

Fig. 7.1 Frequency Histogram of lengths of 150 Rods

Solution 7.8

Since the sample size is $n = 150$, meaning 150 measurements comprise the sample, then the median or centre measurement occurs at the $\left(\dfrac{n}{2}\right) = \left(\dfrac{150}{2}\right) = 75$, the 75th position.

From the histogram, the sum of the first four class frequencies is $5+9+13+18 = 45$. To get to the 75th measurement requires 30 more measurements, all of which come from the fifth class interval, which has a frequency of 44. Thus, the median lies in the fifth class interval, which is therefore the median class.

The median is the x coordinate value (abscissa) through which the vertical line AB passes and which divides the histogram into two equal areas. Since the area of a histogram is equal to frequency, then the area of the given histogram is 150, the total frequency. As the line

AB divides the histogram into two equal areas, then the area of each half is 75, i.e., half the total frequency. For this to be so, the areas ABCD and ABEF must have frequencies 30 and 14 respectively, for the vertical bar DFEC to have a frequency of 44. Thus,

$$CB = \frac{30}{44} CE = \frac{30}{44} (5) = 3.41.$$

Therefore, the median of the histogram is 219.5 + 3.41= 222.91cm.

Example 7.9

Find the median length of a sample of 150 rods using the percent relative cumulative frequency polygon shown below. This polygon is from example 6.2 and is reproduced here as figure 7.2.

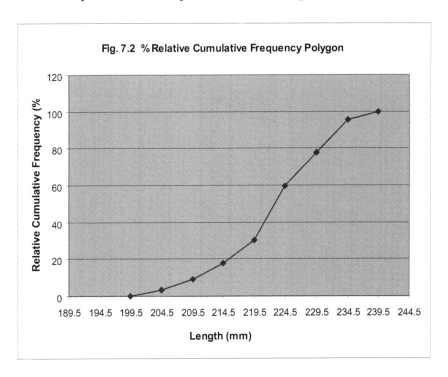

Fig. 7.2 % Relative Cumulative Frequency Polygon

Solution 7.9

From the percent relative cumulative frequency polygon, the median is obtained by drawing a horizontal line through the 50% ordinate (y-axis value) to where it cuts the polygon. From this point, a perpendicular line is drawn to the horizontal axis. The abscissa (x-axis value) at the foot of the perpendicular is the median. From the graph, the median is equal to 222.9 cm.

7.2.3 The Mode

The observation (or measurement) that occurs most frequently in a data set is called the mode. Geometrically, the mode is the highest point on a frequency polygon and the height of the tallest bar on a histogram. If the frequency distribution has only one measurement value that occurs most often, the distribution is said to be **unimodal.** If the distribution has two or more measurement values that occur more frequently than the rest, the distribution is said to be **multimodal.** And if two measurement values occur most often, the distribution is said to be **bimodal.**

Example 7.10

Find the mode for the following data set: 2, 5, 6, 6, 4, 5, 7, 2, 6, 6, 7, 6, 4, 4, 4, 5, 5, 6, 6, 7.

Solution 7.10

Arranging the numbers is ascending order gives
        2, 2, 4, 4, 4, 4, 5, 5, 5, 5, 6, 6, 6, 6, 6, 6, 6, 7, 7, 7

mode = the number that occurs most frequently = 6

Example 7.11

Find the mode for the following set of numbers: 20, 30, 40, 40, 40, 50, 50, 70, 70, 70, 90.

Solution 7.11

Mode = number that occurs most frequently

The numbers 40 and 70 both occur with a frequency of 3. Thus, there are two modes 40 and 70. The data set is bimodal.

7.2.3.1 The Mode for Grouped Data

For grouped data having equal classes, the modal class is the class interval with the largest class frequency.

For a grouped distribution, the mode may be approximated (assuming the distribution is relatively smooth) using the following formula:

$$\text{Mode} = L + \left(\frac{d_1}{d_1 + d_2} \right) W$$

where L is the lower class boundary of the modal class (i.e.,
the class containing the mode)

d_1 is the difference between the frequency of the
modal class and the frequency of the next lower class

d_2 is the difference between the frequency of the
modal class and the frequency of the next higher
class

W is the width of the class interval

Example 7.12

Calculate the modal weight of 26 male students shown in the frequency distribution below.

Weight (kg)	Frequency
51 – 55	2
56 – 60	3
61 – 64	4
65 – 69	7
70 – 74	5
75 – 79	3
80 – 84	2
Total	26

Solution 7.12

From the frequency distribution table, it can be seen that the modal class interval (i.e., the class with the greatest frequency) for the weights is 65–69. Therefore,

$$d_1 = 7 - 4 = 3 \text{ and } d_2 = 7 - 5 = 2$$

and W = 69.5 – 64.5 = 5 (Recall class width
is calculated using class boundaries)

$$\text{mode} = 64.5 + \left(\frac{3}{3+2}\right)(5) = 67.5 \text{ kg}$$

7.2.4. Mean, Median, or Mode: Which Should be Used?

With three types of measures of central tendency available for analysing a data set, choosing the best one of all depends on the type of information one seeks. If the goal is to determine the average size of the values in the data set, then the mean would be the best choice. If the goal is the measure that divides the data set into two equal halves, then the median is the best choice to describe the centre of the distribution. The mode is rarely used as a measure of central tendency because although one may be interested in the measurement that occurs most often, this measurement does not necessarily lie in the centre of the distribution.

The mean is sensitive to very large or very small measurements and tends to shift in the direction of skewness. Thus, for data that is highly skewed, the mean would be a misleading measure of central tendency. On the other hand, the median and the mode are less affected by such skew (i.e., extreme high or low measurements). For skewed distributions, this makes the median a better measure for the centre of the distribution.

7.3 Measures of Dispersion or Variation

Measures of dispersion or variation measure the spread of the measurement values in a data set and hence, the degree of variability among the measurement values in the data set. Measures of dispersion or variation seek to determine whether the values in a data set are clustered close to and around the mean or whether the data values are widely dispersed away from the mean of the data set.

Consider the two illustrations in figure 7.3, both of which are for a sample of 11 cups of 175 ml ice cream taken from two different manufacturing processes. For each process, the mean weight

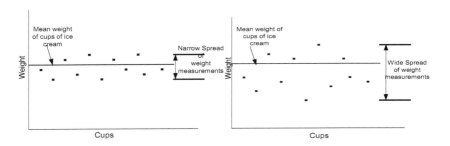

**Fig. 7.3 Weights of 175 ml Cups of Ice Cream
from Two Production Processes**

is the same, but the spread of the weights are considerably less for the process on the left compared to the process on the right. For the process on the left, the weights of the cups of ice cream are close to the mean, whereas for the process on the right, the weights of the cups of ice cream are more widely dispersed away from the mean and also away from each other. Thus, the process on the left produces product with more consistent weights closer to the mean weight and to the weights of each other, implying a more stable process with less variability of product weights compared to the process on the right. The measures of variability that will be considered are

1. Range
2. Standard variation
3. Variance

7.3.1 The Range

The range of a data set is the difference between the largest and smallest measurements in the set and is given by

Range = (largest measurement – smallest measurement)

$$R = x_{max} - x_{min}$$

Example 7.13

Find the range of each data set.
(a) 17, 11, 13, 8, 20, 15, 23, 10
(b) 14, 8, 13, 13, 14, 13, 14, 23

Solution 7.13

(a) largest measurement $x_{max} = 23$

smallest measurement $x_{min} = 8$

Therefore, range R $= x_{max} - x_{min} = 23 - 8 = 15$

(b) largest measurement $x_{max} = 23$

smallest measurement $x_{min} = 8$

Therefore, range R $= x_{max} - x_{min} = 23 - 8 = 15$

In both cases, the range is 15. However, on examination of both data sets, it can be seen that data set (a) has a lot more variation (i.e., different measurement values) than data set (b), which mainly consist of 13s and 14s. This difference in the variability of the two data sets is

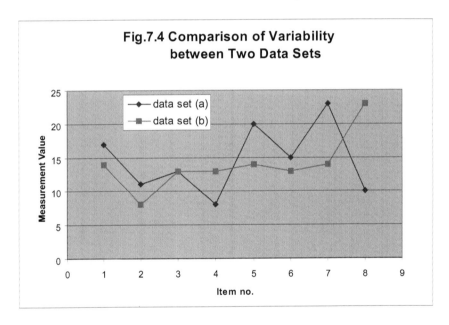

Fig.7.4 Comparison of Variability between Two Data Sets

more strikingly seen by considering a plot of the data as shown in figure 7.4. The value of the range by itself does not reveal that data set (a) has greater variability than data set (b) because it is a function only of the extreme values of a data set. Hence, range is not a very good indicator of dispersion.

7.3.1.1. The Range for Grouped Data

For grouped data, the range may be approximated in one of two ways.

Method 1

Range = (Upper class boundary of the highest class interval) – (Lower class boundary of lowest class interval)

Method 2

Range = (Class mark of the highest class interval) – (Class mark of the lowest class interval)

Example 7.14

Calculate the range of weights of 26 male students shown in the grouped frequency distribution below.

Weight (kg)	Frequency
51 – 55	2
56 – 60	3
61 – 64	4
65 – 69	7
70 – 74	5
75 – 79	3
80 – 84	2
Total	26

Solution 7.14

Method 1

Range = (Upper class boundary of the highest class interval) – (Lower class boundary of lowest class interval)

$= 84.5 - 50.5 = 30.0$ kg

Method 2

Range = (Class mark of the highest class interval) – (Class mark of the lowest class interval)

Now, class midpoint of highest class interval $= \dfrac{80+84}{2} = \dfrac{164}{2} = 82$

and class midpoint of lowest class interval $= \dfrac{51+55}{2} = \dfrac{106}{2} = 53$

Therefore,

$$\text{Range} = 82 - 53 = 29 \text{ kg.}$$

7.3.2 The Standard Deviation

A better measure of dispersion or variability of data is obtained using the numerical descriptive measure called the standard deviation. The calculation of the standard deviation uses *all* the data values and their *distances* from the mean. The larger the standard deviation, the more dispersed the data values are from the mean, whereas the smaller the standard deviation, the closer the data values are to the mean.

The standard deviation of a sample data set is defined by the formula

$$s = \sqrt{\frac{\sum (x - \bar{x})^2}{n-1}}$$

where s is the sample standard deviation
x is a measurement value
$\bar{x}$ is the mean value of the data set

n is the sample size or number of measurements comprising the sample

Let's take a closer look at this formula. Consider the numerator. The term $x - \bar{x}$ measures how far away each data value is from the mean and is called the **deviation**. After calculating the size of the deviation for each data value, it is then squared and summed. The result is then divided by $n-1$ and the square root taken. Thus, the standard deviation is the square root of the sum of the squares of the deviations of the measurements divided by $n-1$. The square of the sample standard deviation is called the **variance** and is denoted by s^2.

An alternative formula that is computationally simpler to use for calculating standard deviation is given by

$$s = \sqrt{\frac{\sum x^2 - \frac{(\sum x)^2}{n}}{n-1}}$$

Example 7.15

Calculate the standard deviation for each data set.
   (a) 17, 11, 13, 5, 20, 15, 23, 8
   (b) 14, 8, 13, 13, 14, 13, 14, 23

Solution 7.15

(a) Using the definitional formula

Mean $\bar{x} = \dfrac{\sum x}{n} = \dfrac{17 + 11 + 13 + 5 + 20 + 15 + 23 + 8}{8} = \dfrac{112}{8} = 14$

The solution is set out in tabular form below:

x	$x-\bar{x}=x-14$	$(x-\bar{x})^2$
17	3	9
11	-3	9
13	-1	1
5	-9	81
20	6	36
15	1	1
23	9	81
8	-6	36

$$\sum x = 112 \qquad\qquad \sum (x-\bar{x})^2 = 254$$

$$s = \sqrt{\frac{\sum (x-\bar{x})^2}{n-1}}$$

$$= \sqrt{\frac{254}{8-1}} = \sqrt{\frac{254}{7}} = \sqrt{36.28}$$

$$= 6.02$$

(b) Using the computational formula

x	x^2
14	196
8	64
13	169
13	169
14	196
13	169
14	196
23	529

$$\sum x = 112 \qquad \sum x^2 = 1688$$

Using

$$s = \sqrt{\frac{\sum x^2 - \frac{(\sum x)^2}{n}}{n-1}}$$

$$= \sqrt{\frac{1688 - \frac{(112)^2}{8}}{8-1}}$$

$$= \sqrt{\frac{1688 - 1568}{7}} = \sqrt{17.14}$$

$$= 4.14$$

Notice that the two data sets have the same size and mean, that is, 8 measurements comprise each data set, and the mean of each data set is 14. However, the values of data set (a) are much farther apart from each other and are therefore farther from the mean value than do the values of data set (b). Thus, data set (a) has more variability than data set (b), which is shown by the standard deviations. Data set (b) has a standard deviation of 4.14, which is smaller than that of data set (a), which has a standard deviation of 6.02. Smaller standard deviation means less variability among the data values. The practical importance of being able to assess the variability of data sets is used in production processes, where small variability of some characteristic of a product being produced may indicate more consistent quality. Recall the case of the weights of the 175 ml cups of ice cream illustrated in figure 7.3.

Example 7.16

Calculate the standard deviation of the following frequency distribution.

Measurement (x)	14	12	9	8	7
Frequency (f)	2	5	7	4	1

Solution 7.16

The definitional formula for the sample standard deviation for a frequency distribution is given by

$$s = \sqrt{\frac{\Sigma f(x - \bar{x})^2}{n-1}}$$

Now, the mean is

$$\bar{x} = \frac{\Sigma xf}{\Sigma f}$$

$$= \frac{(14)(2)+(12)(5)+(9)(7)+(8)(4)+(7)(1)}{2+5+7+4+1}$$

$$= \frac{28+60+63+32+7}{19} = \frac{190}{19}$$

$$= 10$$

$$\Rightarrow \bar{x} = 10$$

The solution is set out in tabular form below:

x	f	$x-\bar{x}=x-10$	$(x-\bar{x})^2$	$f(x-\bar{x})^2$
14	2	14-10= 4	$4^2=16$	2(16)= 32
12	5	12-10= 2	$2^2=4$	5(4) = 20
9	7	9-10= -1	$-1^2=1$	7(1) = 7
8	4	8-10= -2	$-2^2=4$	4(4) = 16
7	1	7-10= -3	$-3^2=9$	1(9) = 9
				$\sum f(x-\bar{x})^2=84$

Therefore,

$$s = \sqrt{\frac{84}{19-1}} = \sqrt{\frac{84}{18}} = \sqrt{4.66} = 2.1$$

The computational formula for the sample standard deviation of a frequency distribution is given by

$$s = \sqrt{\frac{\sum fx^2 - \frac{(\sum fx)^2}{n}}{n-1}}$$

The solution is set out in tabular form below:

x	f	fx	x^2	fx^2
14	2	2(14)= 28	196	2(196)= 392
12	5	5(12)= 60	144	5(144) = 720
9	7	7(9)= 63	81	7(81) = 567
8	4	4(8)= 32	64	4(64) = 256
7	1	1(7) = 7	49	1(49) = 49
		$\sum fx = 190$		$\sum fx^2 = 1984$

Substituting in the computational formula gives

$$s = \sqrt{\dfrac{1984 - \dfrac{(190)^2}{19}}{19-1}} = \sqrt{\dfrac{(1984-1900)}{18}}$$

$$= \sqrt{\dfrac{84}{18}} = \sqrt{4.66} = 2.1$$

Example 7.17

Calculate the standard deviation for the sample weights of 26 male students shown in the grouped frequency distribution below.

Weight (kg)	Frequency
51 – 55	2
56 – 60	3
61 – 64	4
65 – 69	7
70 – 74	5
75 – 79	3
80 - 84	2
Total	26

Solution 7.18

For grouped data, recall that the individuality of the raw data is lost and as such, it is assumed that the individual observations (measurements) in a given class interval lie at the centre of the class interval, called the class mark (or class midpoint) x_m. The standard deviation definitional formula for grouped data is given by

$$S = \sqrt{\frac{\Sigma f (x_m - \bar{x})^2}{n-1}}$$

and the computationally simpler formula is given by

$$S = \sqrt{\frac{\Sigma f x_m^2 - \dfrac{(\Sigma f x_m)^2}{n}}{n-1}}$$

For a grouped distribution the mean of the distribution is given by

$$\text{Mean } \bar{x} = \frac{\Sigma x_m f}{\Sigma f}$$

The class midpoint, as you will remember, is the average of the lower and upper class limits. The table below shows the original table with the class midpoints included.

Weight class (kg)	Class midpoint (x_m)	Frequency (f)	$x_m f$	$x_m - \bar{x} =$ $x_m - 67.5$	$(x_m - \bar{x})^2$	$f(x_m - \bar{x})^2$
51-55	53	2	106	53.0-67.5= -14.5	210.25	2(210.25)= 410.50
56-60	58	3	174	58.0-67.5 = -9.5	90.25	3(90.25)= 270.75
61-64	63	4	252	63.0-67.5=-4.5	20.25	4(20.25)= 81.00
65-69	67	7	469	67.0-67.5=-0.5	0.25	7(0.25)= 1.75
70-74	72	5	360	72.0-67.5=4.5	20.25	5(20.25)= 101.25
75-79	77	3	231	77.0-67.5=9.5	90.25	3(90.25)= 270.75
80-84	82	2	164	82.0-67.5=14.5	210.25	2(210.25)= 410.50

$$\Sigma f = 26 \qquad \Sigma x_m f = 1756$$

$$\Sigma f(x_m - \bar{x})^2 = 1546.50$$

The mean $\bar{x} = \dfrac{\Sigma x_m f}{\Sigma f} = \dfrac{1756}{26} = 67.5 \text{ kg}$

Substituting in the formula

$$s = \sqrt{\frac{\Sigma f(x_m - \bar{x})}{n-1}}$$

$$= \sqrt{\frac{1546.5}{26-1}} = \sqrt{\frac{1546.5}{25}}$$

$$= \sqrt{61.86} = 7.9\,kg$$

The solution is now presented using the computationally simpler formula

$$s = \sqrt{\frac{\Sigma fx_m^2 - \dfrac{(\Sigma fx_m)^2}{n}}{n-1}}$$

The solution to the formula is shown below in tabular form.

Weight Class (kg)	Class Midpoint (x_m)	Freq (f)	x_m^2	fx_m	fx_m^2
51-55	53	2	2809	2(53)= 106	2(2809)= 5618
56-60	58	3	3364	3(58)= 174	3(3364)= 10092
61-64	63	4	3969	4(63)= 252	4(3969)=15876
65-69	67	7	4489	7(67)= 469	7(4489)= 31423
70-74	72	5	5184	5(72)= 360	5(5184)= 25920
75-79	77	3	5929	3(77)= 231	3(5929)=17787
80-84	82	2	6724	2(82)= 164	2(6724)= 13448

$$\Sigma f = 26 \qquad \Sigma fx_m = 1756 \qquad \Sigma fx_m^2 = 120,164$$

Substituting into

$$s = \sqrt{\frac{\Sigma f x_m^2 - \dfrac{\left(\Sigma f x_m\right)^2}{n}}{n-1}}$$

$$= \sqrt{\frac{120,164 - \dfrac{(1756)^2}{26}}{26-1}}$$

$$= \sqrt{\frac{120,164 - 118,597.5}{25}} = \sqrt{\frac{1566.5}{25}}$$

$$= \sqrt{62.6} = 7.9\,\text{kg}$$

7.3.2.1 Relating Standard Deviation to Data Set Values

If a data set forms a normal curve, the mean and standard deviation for that data set can be used together to arrive at three intervals into which most of the measurements in the data set can be placed. These intervals are expressed in terms of *the number of standard deviations* on either side of the mean and are defined as follows:

1st interval. Called one standard deviation of the mean and is equal to $\bar{x} \pm s$

2nd interval. Called two standard deviations of the mean and is equal to $\bar{x} \pm 2s$

3rd Interval. Called three standard deviations of the mean, and is equal to $\bar{x} \pm 3s$

These intervals are shown in figure 7.5 for a data set, which when plotted has a curve that is bell-shaped and is symmetrical.

The number of measurements and hence the proportion or percentage of the total number of measurements in the data set falling within each of the three intervals can then be determined.

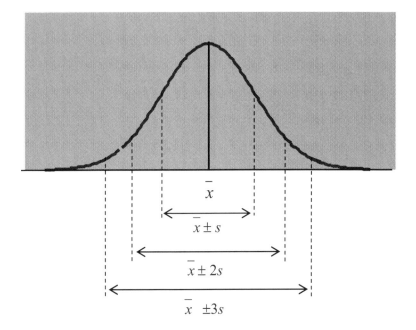

Fig. 7.5 Intervals Showing No. of Standard
Deviations About the Mean

It has been established that for data sets that form a bell shaped or normal curve that

1. 68.3% of all the measurements in the data set lie within 1 standard deviation on either side of the mean, i.e., between $\bar{x} - s$ and $\bar{x} + s$.

2. 95.5% of all the measurements in the data set lie within 2 standard deviations on either side of the mean, i.e., between $\bar{x} - 2s$ and $\bar{x} + 2s$.

3. 99.7% of all the measurements in the data set lie within 3 standard deviations on either side of the mean, i.e., between $\bar{x} - 3s$ and $\bar{x} + 3s$.

Figure 7.6 shows these percentage values, which also correspond to the area under the normal or bell-shaped curve.

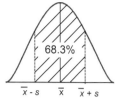 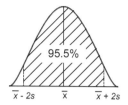 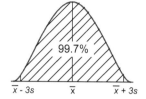

(a) 1 standard deviation   (b) 2 standard deviations   (c) 3 standard deviations
about the mean   about the mean   about the mean

Fig. 7.6 No. of Standard Deviations and Corresponding Area under the Normal Curve Expressed as a Percentage of the Total Area Under the Curve

7.4 Measures of Relative Standing

Measures of relative standing describe the position of an individual observation (measurement) in a data set or distribution. That is, it describes the position of a measurement in relation to the other measurements comprising the data set or distribution. As an example, consider a data set made up of the daily allowance each child receives in a class of 25 students. Suppose one child reports that he/she receives a daily allowance of $8.00. We might want to know if this is a relatively high or a relatively low allowance. What percentage of daily allowances of students in the class is less than $8.00, and what percentage is higher? In other words, where does $8.00 lie with respect to all the 25 values making up the allowance data set? Is it at the lower end, somewhere in the middle, or towards the upper end? Answers to these questions, which describe the relative position of a particular measurement in relation to the other measurements, are called measures of relative standing.

Three methods used to describe measures of relative standing are

- Percentiles
- Quartiles
- z-scores

When describing the relative standing of measurements, it is customary to refer to the measurements as scores.

7.4.1 Percentile

A **percentile** is a measure used to describe the position of a measurement in a data set in terms of a percentage. Consider a data set made up of 10 measurements arranged in ascending order: $x_1, x_2, x_3, \ldots \ldots x_{10}$. Since we are dealing with percentiles, measurements are called scores. The percentile rank of the score x_7 is the percentage of scores that are less than or equal to x_7.

When using the percentile measure to describe relative standing, two questions may be asked about the position of a measurement (score) in a data set:

1. What percentage of measurements (scores) are equal to or less than a particular score (measurement), ?
2. What score (measurement value) in a data set corresponds to a given percentile rank?

Example 7.19

The following marks were obtained by 16 students in a statistics examination:

19, 27, 33, 34, 35, 37, 41, 44, 45, 47, 48, 50, 54, 57, 58, 59

(a) What is the percentile rank of the student who scored 41 marks?

(b) A student is ranked in the 88 percentile. What mark does this correspond to?

Solution 7.19

(a) The percentile rank of the student who scored 41 marks is the percentage of marks received that are less than or equal to 41.

The marks less than or equal to 41 are 19, 27, 33, 34, 35, 37, 41. There are 7 of them.

So the percentile rank of the student who scored 41 is $\frac{7}{16} \times 100$ % = 43.75%

This is interpreted to mean that 43.75% of the scores received are equal to or less than 41.

(b) A student who is ranked in the 88th percentile means that 88% of the students (or 88% of the marks in the data set) fall below the 88th percentile. That is, 88% of the marks in the data set are less than or equal to the mark x corresponding to the 88th percentile. Thus,

$$\frac{\text{No. of marks} \leq x}{16} = 0.88$$

No. of marks $\leq x = (0.88)\,(16) = 14$ to the nearest integer

Thus, the 14th mark in the data set corresponds to the mark at the 88th percentile. The 14th mark is 57, and therefore, the student who ranked in the 88th percentile scored 57 marks.

7.4.2 Quartiles

Just as percentiles divide a data set into 100 equal parts (percent means per hundred), quartiles divide a data set into four equal parts, with each part containing one fourth or 25% of the measurements (observations).

Consider the diagram below in figure 7.7.

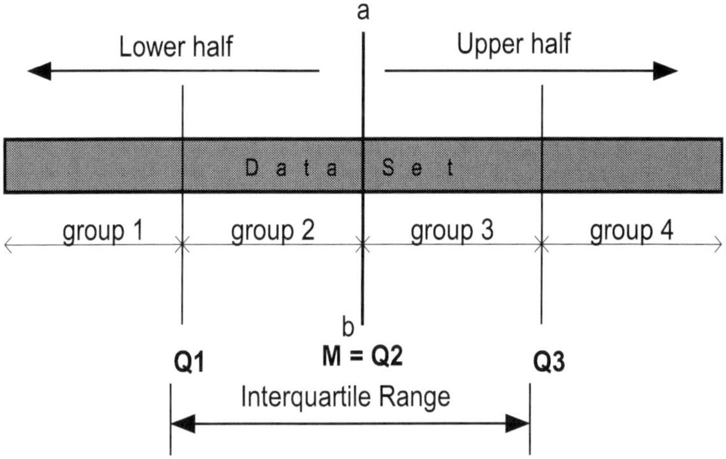

Fig.7.7 Illustration of Quartiles

The shaded horizontal rectangular bar represents a data set arranged in increasing order from left to right. The vertical line *ab* locates the median M of the data set. Recall that the median M divides the data set into two equal halves. That to the left of *ab* is the lower half of the data set, and that to the right is the upper half of the data set.

The 1st, or lower, quartile (Q1), divides the lower half of the data set into two equal parts and may be considered the median of the lower half of the data set. The 2nd, or mid, quartile (Q2) is in fact the median of the entire data set. And the 3rd, or upper, quartile (Q3) divides the upper half of the data set into two equal parts and may also be considered the median of the upper half of the data set.

Thus, the quartiles Q1, Q2, and Q3 divide the entire data set into four equal parts or groups as shown in figure 7.7. It should be clear that the first quartile is also the 25th percentile, the 2nd or mid quartile is the 50th percentile, and the 3rd quartile is the 75th percentile.

The **interquartile range (IQR)** is the distance between the 1st and 3rd quartiles and is given by IQR = Q3 – Q1 and contains 50% of the measurements in the data set. The larger the interquartile range, the greater the variability in the data.

For a small data set, meaning one that contains 35 or less measurements, the following are defined:

The lower quartile Q1 is located at the ¼ $(n + 1)$ position.
The mid quartile Q2 is located at the ½ $(n + 1)$ position.
The upper quartile Q3 is located at the ¾ $(n + 1)$ position.

In this case, n is the number of members in the data set. For large data sets—those that contain more than 35 members—the +1 is omitted from the above definitions.

Example 7.20

Consider the following sample of 20 measurements:

| 29 | 35 | 22 | 34 | 34 | 23 | 14 | 28 | 39 | 37 |
| 40 | 18 | 38 | 31 | 30 | 24 | 27 | 18 | 36 | 27 |

Find and interpret

   (i)   the lower quartile, Q1
   (ii) the median, M
   (iii) the upper quartile, Q3
   (iv) the semi-interquartile range
   (v) the 88th percentile for the data set

Solution 7.20

First, arrange the measurements in ascending order.

14, 18, 18, 22, 23, 24, 27, 27, 28, 29, 30,
31, 34, 34, 35, 36, 37, 38, 39, 40

(i) The lower quartile Q1 is located at the ¼ (n+1) position. That is, ¼(20+1) = 5.25 ≅ 5 to the nearest integer. The number at the 5th position is 23.

Therefore, the lower quartile Q1 = 23. This means that 25% of the measurements fall below the 1st quartile, 23.

(ii) Since the sample size is 20, an even number, the median is the arithmetic mean of the two middle numbers located at the

$$\left(\frac{n}{2}\right) \text{ and } \left(\frac{n}{2}+1\right) \text{ positions, i.e., } 10^{th} \text{ and } 11^{th} \text{ positions.}$$

Therefore, median = $\left(\frac{28+29}{2}\right)$ = 28.5. This means that 50% of the measurements fall below the median, 28.5.

(iii) The upper quartile Q3 is located at the ¾(n+1) position. That is, ¾(20+1) = 15.75 ≅ 16 to the nearest integer. The number at the 16th position is 36.

Therefore, the upper quartile Q3 = 36. This means that 75% of the measurements fall below the 3rd quartile, 36.

(iv) The semi-interquartile = ½(Q3−Q1) = ½(36−23) = 6.5

The semi interquartile range is a measure of the spread of the measurements centred around the median and only covers half of the data set. The smaller its value, the less spread the data has between the quartiles.

(v) The 88th percentile is a score in the data set such that 88% of the measurements in the data set fall below the 88th percentile. Thus, the 88th percentile is the measurement value located at the (0.88)(20) = 17.6 or 17th position. This measurement value is 37. Thus, 88% of the measurements in the data set fall below the value 37.

7.4.3 z-Scores

Recall that if a plot of the measurements values vs. frequency for a variable results in a bell-shaped curve, the curve is called a normal curve, and the variable under investigation is called a normal variable. Alternatively, the data set (frequency distribution table) comprising the measurement values is referred to as a normal distribution.

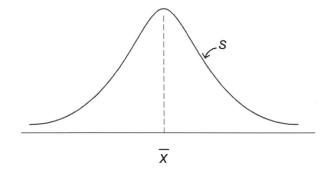

Fig. 7.8 Normal Curve

Figure 7.8 shows a normal curve having a sample mean $\overline{x}$ and sample standard deviation s. Since the normal curve is symmetrical, the axis of symmetry is centred on the mean, $\overline{x}$. The sample standard deviation determines how "spread out" the curve is. The larger the value of s, the more spread out is the curve from $\overline{x}$.

Figure 7.9 shows normal curves having different means and standard deviations. It should be noticed that although all normal curves have a bell shape, their appearance are determined by the mean and standard deviation. Figure 7.9a shows two normal curves having the same amount of spread and hence the same standard deviation but different means. Different means cause the curves to shift along the x-axis. Figure 7.9b shows two normal curves having the same mean but different standard deviations. The curve with the smaller standard deviation results in that variable under investigation having measurement values that are closer to the mean than those of the variable with standard deviation $s=7$. Hence, the curve with the smaller standard deviation has a taller peak and lower tails. Figure 7.9c shows two normal curves with different means and different standard deviations.

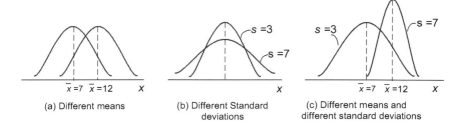

(a) Different means   (b) Different Standard deviations   (c) Different means and different standard deviations

Fig. 7.9 Appearance of Normal Curves

Since different normal curves will have different means and standard deviations, it is more convenient to construct a **standardized normal curve** for all variables that have a normal distribution. This standardized normal curve has a mean of zero and a standard deviation of 1. The area under the standardized normal curve is always equal to 1, and since the curve is symmetrical about the mean, the area under the curve to the left or right of the mean is ½. The normal variable x is transformed into a standardized normal variable z, called a z-score, which is defined as

$$z = \frac{x - \bar{x}}{s}$$

Consider the right hand side (RHS) of the equation. The term $x - \bar{x}$ gives the distance of x from $\bar{x}$. If $x - \bar{x}$ is positive, then x lies to the right of $\bar{x}$. If $x - \bar{x}$ is negative, then x lies to the left of $\bar{x}$. Dividing the result of $x - \bar{x}$ by s expresses this deviation of x from $\bar{x}$ in units of number of standard deviations. Thus, the z-score of a measurement x is the distance of the measurement x above or below the mean $\bar{x}$ of the data set, expressed in units of standard deviations.

A negative z-score indicates that the observation (measurement) lies to the left of the mean, whereas a positive z-score indicates that the measurement lies to the right of the mean.

Example 7.21

Compute the z-score of a normal variable $x = 7$ that lies in a data set that has a mean of 4 and a standard deviation of 2.

Solution 7.21

It is given that $\bar{x} = 4$ and $s = 2$.
Using

$$z = \frac{x - \bar{x}}{s}$$

$$= \frac{7 - 4}{2} = \frac{3}{2} = 1.5$$

$$\Rightarrow z = 1.5$$

This is interpreted to mean that the measurement $x = 7$ lies a distance of 1.5 standard deviation units to the right of the mean $= 4$.

Had the problem been to determine the z-score of $x = 1$, the answer would have been

$z = -1.5$. This would now be interpreted to mean that the measurement $x = 1$ lies a distance of 1.5 standard deviation units to the left (since z is negative) of the mean $= 4$.

7.5 Measures of Association

The numerical descriptive measures discussed in this chapter thus far were applied to data sets or frequency distribution tables in which the experimental units were subjected to one set of measurements. If the experimental units under investigation are subjected to two sets of measurements (e.g., the height and weight measurements of a group of students in a class), we might be interested in determining if some type of relationship exists between these two sets of measurements. The numerical descriptive measure that is used in determining whether a relationship exists between two sets of measurements made on the same experimental units is called measures of association.

Consider a group of experimental units on which we want to investigate the relationship between two characteristics among them. Let the two characteristics be denoted by the variables X and Y. For each experimental unit, numerical measurements x and y are taken and recorded for both X and Y respectively. Each pair of numbers for X and Y are then plotted on a rectangular coordinate system to form a scatter plot. The scatter plot gives a visual insight into the relationship between X and Y. Figure 7.10 shows some possible scatter plot patterns for X and Y.

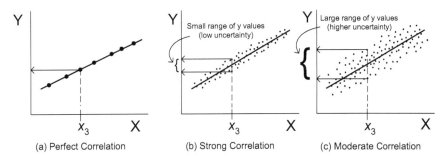

(a) Perfect Correlation   (b) Strong Correlation   (c) Moderate Correlation

Fig. 7.10 Scatter Plot Patterns Showing
Linear Positive Correlation
for Variables X and Y

If all the (x, y) points lie exactly along a straight line as in figure 7.10a, then for any value of x, there is one and only one value of y, and we say that the two variables X and Y show a perfect linear relationship or correlation. Since increasing values of x results in increasing values of y, the linear correlation is said to be positive or direct. However, if a given value of x results in a range of possible values of y as shown in figure 7.10b and 7.10c, then there is a degree of uncertainty about the value of y. This degree of uncertainty for a given x value, say x_3, is smaller for the scatter plot of figure 7.10b than for figure 7.10c. This is because the spread of the points about the superimposed straight line (for perfect correlation) in figure 7.10b is narrower than in figure 7.10c. Since the spread of the points about the straight line and hence the range of possible y values is smaller for figure 7.10b than for figure 7.10c, we say that the former plot shows a stronger linear relationship or a stronger correlation between the X and Y variables. Thus, figure 7.10b may be described as having strong linear correlation between the X and Y variables and figure 7.10c as having moderate linear correlation between the X and Y variables. In general, the narrower the range of variation of y values for a given value of x, the more strongly the two variables X and Y are correlated.

The scatter plots of figure 7.10 are for two variables that show positive or direct linear correlation. Figure 7.11 show scatter plots for two variables that show negative or inverse linear correlation. With negative correlation, Y decreases as X increases, and the same principles discussed above for linear positive correlation holds for linear negative correlation.

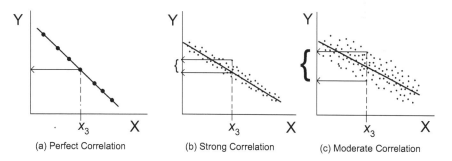

(a) Perfect Correlation   (b) Strong Correlation   (c) Moderate Correlation

Fig. 7.11 Scatter Plot Patterns Showing
Linear Negative Correlation
for Variables X and Y

If the scatter plot for the (x, y) points reveal a pattern as shown in figure 7.12, where the range of y values for a given x value, x_3, is almost the same as the range of y values for the entire group, then there is no relation and hence no correlation between the X and Y variables.

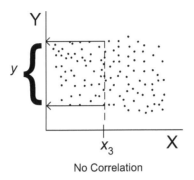

No Correlation

Fig. 7.12 Scatter Plot Pattern Showing No Correlation Between Variables X and Y

7.5.1 The Coefficient of Correlation

The strength and directionality of a linear relationship between two variables may be expressed quantitatively in terms of a correlation coefficient denoted by r. The correlation coefficient is defined as:

$$r = \frac{n\Sigma xy - \Sigma x \Sigma y}{\sqrt{[n\Sigma x^2 - (\Sigma x)^2][n\Sigma y^2 - (\Sigma y)^2]}}$$

where x and y are the measurement values and n is the number of x and y measurements.

The value of r can lie between +1 and −1, inclusive. The numerical value of r indicates the degree or strength of the linear relationship between the two variables. The closer the numerical value of r to unity, the stronger is the linear relationship between the two variables. If the sign of r is positive, the linear relationship between the two variables is direct, meaning as the value of x increases, the value of y also increases. If the sign of r is negative, the linear relationship is described as being inversely related, meaning as the value of

x increases, the value of y decreases. If r is equal to +1 or −1, the two variables show perfect positive correlation or perfect negative correlation respectively, and any given value of x is associated with one and only one value of y. The closer the value of r lies to zero means that there is little linear correlation between the values of x and y, there being no linear correlation when r is equal to zero.

It should be noted, that although no linear correlation may exist between two variables, a nonlinear or curvilinear relationship may exist. However, this type of relationship is not discussed in this text.

In summary, if information is gathered about two variables for a group of experimental units, we can assess whether these two variables show some type of linear relationship by doing a scatter plot of the two variables. If the scatter plot results in a straight line, then a perfect linear correlation exists between the two variables. If the scatter plot clusters about a best-fit straight line drawn through the data points, then there exists a relationship between the two variables. If the data points form a narrow band centred on the straight line, then for a given x value, there is a narrow range of y values that can be associated with that x value. However, if the data points form a wide band centred on the straight line, then for a given x value, there is a wider range of y values that may be associated with that x value. In this case, we say that there is a greater uncertainty in the value of y associated with a given x value. Establishing how well two variables are correlated enables one to make a prediction about one variable, given the other variable. The accuracy of the prediction is related to the degree or strength of the linear relationship between the two variables.

Example 7.22

The data set below shows the values for two variables, X and Y.

X	4	5	6	7	4	5	6	7	4	5	6	7
Y	11	10	9	8	10	9	8	7	9	8	7	6

(a) Plot the scatter diagram.
(b) State whether or not the two values show a linear relationship and if so, whether the relationship is positive or negative.
(c) Determine the coefficient of correlation, r.

Solution 7.22

(a) The scatter plot is shown in the figure below.

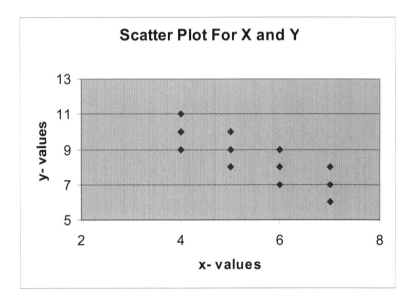

(c) The two values show a negative linear relationship. Observe that in general, as the value of x increases, the value of y decreases.

(d) The coefficient of correlation, r, is calculated using the formula below.

$$r = \frac{n\Sigma xy - \Sigma x \Sigma y}{\sqrt{[n\Sigma x^2 - (\Sigma x)^2][n\Sigma y^2 - (\Sigma y)^2]}}$$

where $n = 12$, the number of observations for x and y.

As an aid in evaluating this formula, the calculation is set out in tabular form below.

x	y	x^2	y^2	xy
4	11	16	121	44
5	10	25	100	50
6	9	36	81	54
7	8	49	64	56
4	10	16	100	40
5	9	25	81	45
6	8	36	64	48
7	7	49	49	49
4	9	16	81	36
5	8	25	64	40
6	7	36	49	42
7	6	49	36	42

$\Sigma x = 66$ | $\Sigma y = 102$ | $\Sigma x^2 = 378$ | $\Sigma y^2 = 890$ | $\Sigma xy = 546$

Substituting in the formula for r gives

$$r = \frac{12(546) - (66)(102)}{\sqrt{[12(378) - (66)^2][12(890) - (102)^2]}}$$

$$= \frac{6552 - 6732}{\sqrt{[4536 - 4356][10680 - 10404]}}$$

$$= \frac{-180}{\sqrt{[180][276]}}$$

$$= \frac{-180}{\sqrt{49680}} = \frac{-180}{222.9} = -0.8075$$

$$\Rightarrow r = -0.8075$$

The value of r shows that there is fairly high linear correlation between the two variables.

In closing, if the correlation coefficient r for two variables have a large positive or negative value, the only conclusion that may be drawn is that there exists a linear relationship between those variables. We cannot conclude that a change in x causes the change in y, for such a conclusion means that there is a *causal* relationship between the two variables. Causal meaning that the change in the dependent variable y is due directly to the change in the independent variable x. But there may be other factors that also contribute to the change in the dependent variable y. Keep in mind that a high value of r (strong correlation) does not imply causality.

Statistical Terms

descriptive statistics. This refers to numerical quantities that summarise the important characteristics about a data set. These numerical quantities that summarise the characteristics of the data set are calculated using the numbers that comprise the data set.

measures of central tendency. These are numbers that locate where most of the measurements in the data set tend to cluster or try to locate the centre of the data set

measures of dispersion. These are numbers that measure the amount of variability among the measurements comprising the data set

measures of relative standing. These are numbers used to determine where a particular measurement in a data set lies in relation to all the other measurements in the data set

measures of association. This is a number used to determine if a relationship or correlation exists between two different variables

mean. The arithmetic average of the measurements making up the data set

median. A number that divides a data set into two equal halves after the measurements in the data set are arranged in either ascending or descending order

median class. When data is arranged in a grouped format, the median class is the class interval in which the median measurement lies.

mode. The most frequent measurement occurring in a data set

range. The difference between the largest and smallest measurements in a data set

deviation. This is a measure of how far away a given measurement in a data set is from the mean value of the measurements of the data set. Quantitatively, it is the difference between the mean value of a data set and any given measurement in the data set.

percentile. A measure used to describe the position of a measurement in a data set in terms of a percentage

quartile. A measure used to divide a data set into four equal parts

variance. The square of the standard deviation

standardized normal curve. A normal or bell-shaped curve that has a mean of zero and a standard deviation of 1

coefficient of correlation. A number that determines the strength of the linear relationship between two variables. It is denoted by r and has a value between between −1 and +1.

normal variable. A variable that can take on values that follow a bell-shaped curve

CHAPTER **8**

Linear Regression

8.1 Introduction

In the last section of the previous chapter, we looked at scatter plots between two variables denoted by X and Y. The scatter plots considered showed some general linear relationship, and using the data points that made up the scatter plot, we calculated a number—called the coefficient of correlation r—that told us how well the two variables obeyed the linear relationship. In this chapter, we will look at some methods of fitting the best straight line to the scatter plot data points. The first two methods discussed are subjective and dependent on the observer (i.e., the person fitting the straight line). The third method, called **linear regression**, is rigorous and is a statistical technique used to generate an equation of a straight line that best fits or describes the scatter data plot. With the straight line fitted to the data points, it will then be used to predict the value of one variable given the value of the other variable.

8.2 Coordinate Geometry of the Straight Line

Before getting into the details of fitting straight lines to data points, the basic mathematics about the straight line will be reviewed.

8.2.1 Gradient of a Straight Line

Figure 8.1 shows a rectangular coordinate axis with three straight lines labeled L_1, L_2, and L_3. Notice that L_1 and L_2 both rise upward from left to right, but line L_1 is steeper than that of L_2. We say that the slope or gradient of L_1 is greater than that of L_2. Now look at the straight line L_3. This line slopes downward from left to right. Lines L_1 and L_2 are said to have a positive slope and line L_3 a negative slope.

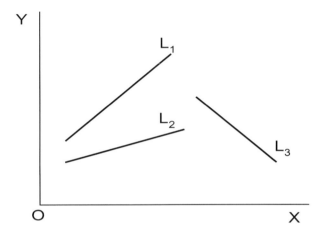

Fig. 8.1 Direction of Slope of Straight Lines

Consider a straight line AB passing through the points $P(x_1, y_1)$ and $Q(x_2, y_2)$ as shown in figure 8.2.

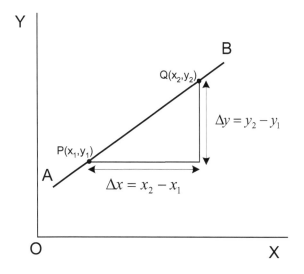

Fig. 8.2 Finding the Slope of a Straight Line

A quantitative value for the slope or gradient (denoted by *m*) of the straight line AB is defined as

$$\text{Slope or Gradient, } m = \frac{\text{change in } y}{\text{change in } x}$$

$$= \frac{\Delta y}{\Delta x} = \frac{y_2 - y_1}{x_2 - x_1}$$

Example 8.1

Find the gradient of the straight line passing through the points
        (i) (-3, -4) and (4, 6)
        (ii) (-3, 6) and (5, -4)

Solution 8.1

(i) Let $(x_1, y_1) = (-3, -4)$ and $(x_2, y_2) = (4, 6)$

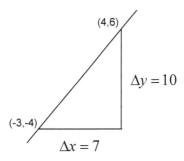

$$\text{Gradient } m = \frac{y_2 - y_1}{x_2 - x_1} = \frac{6 - (-4)}{4 - (-3)} = \frac{6 + 4}{4 + 3} = \frac{10}{7}$$

$$\Rightarrow m = \frac{10}{7}, \text{ a positive gradient}$$

(ii) Let $(x_1, y_1) = (-3, 6)$ and $(x_2, y_2) = (5, -4)$

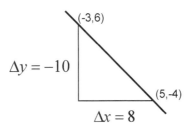

$$\text{Slope } m = \frac{y_2 - y_1}{x_2 - x_1} = \frac{-4 - 6}{5 - (-3)} = \frac{-4 - 6}{5 + 3} = \frac{-10}{8} = \frac{-5}{4}$$

$$m = -\frac{5}{4}, \text{ a negative slope}$$

8.3 Equation of a Straight Line

Given two variables x and y where x is the independent variable and y is the dependent variable, then if x and y are linearly related, they may be represented by an equation of the form

$$y = mx + c$$

where *m* and *c* are constants. *m* is the slope or gradient of the straight line, and *c* is called the *y*-intercept. It is the point at which the straight line cuts the y-axis. Note that if *c* is equal to zero, the equation of the straight line reduces to

$$y = mx$$

and the straight passes through the origin, O, that is, the point (0, 0). These two general equations are represented graphically below in figure 8.3.

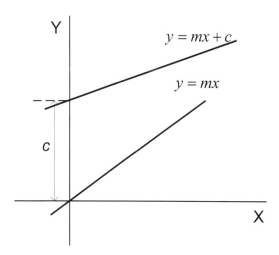

Fig. 8.3 Equations of Straight Lines

Consider the equation $y = mx$. If the slope is 1, then the equation becomes $y=x$. This means that whatever the value of x, y has the same value. Thus, the points (0, 0), (1, 1), (2, 2), (3, 3), (-1, -1), and (-2, -2) all lie on the line $y=x$ and is said to satisfy the equation $y=x$. Similarly, the straight line having the equation, say $y=3x$, means that whatever the value of the x coordinate, the value of y is three times that value. Thus, the points (1, 3), (2, 6), (3, 9), and (-2, -6) lies on the line $y=3x$ or satisfies the equation $y=3x$.

Consider the straight line (linear) equation $y=2x+5$. The straight line represented by this equation has a slope $m=2$ and a y-intercept

$c=5$. That is, the straight line cuts the Y-axis at a distance 5 units above the origin. The points (-2, -1), (-1, 3), (0, 5), (1, 7), (2, 9), and (3, 11) all lie on the line $y=2x+5$ or satisfy the equation $y=2x+5$. The general equation, $y=mx+c$, is called the **slope-intercept** form of the equation of a straight line.

If a straight line has a slope m and passes through a point (h, k), the equation of the straight line is given by

$$y-k=m(x-h)$$

This form of the straight line equation is called the **point-slope** form because the slope m of the line is known as well as a point (h, k) through which the line passes.

Still another form of the equation of a straight line, given by

$$ax+by+c=0$$

is called the **general linear** form of the equation of a straight line. Any straight line equation can be expressed in any one of these forms, and given a straight line equation in any one of these forms, it may be expressed in any of the other two forms.

Example 8.2

(a) Find the equation of the line that has a slope of 3 and passes through the point (4, 7). Sketch the line.
(b) Find
    (i) y when $x = 2$ and
    (ii) x when $y =8$

Solution 8.2

(a) In this problem, we know the slope $m=3$ and a point through which the line passes. Let this point be $(h, k) = (4, 7)$.

Therefore, using the point-slope form of the straight line,

$$y-k=m(x-h)$$

$$\Rightarrow y-7=3(x-4)$$
$$\Rightarrow y-7=3x-12$$
$$\Rightarrow y=3x-12+7$$
$$\Rightarrow y=3x+5$$

To sketch the straight line, we need to know only two points, as two points define a straight line. We already know one point—i.e., (4, 7). The other point may be found by finding the value of y when $x = 0$. Thus, when $x = 0$,

$$y = 3(0) + 5 = 5$$

Thus, the other point is (0, 5). The sketch of the line is shown below.

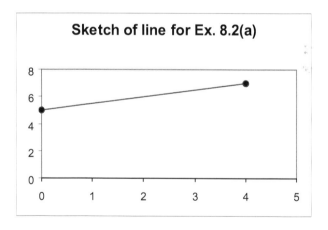

(b) (i) when $x = 2$,

$$\Rightarrow y = 3(2) + 5 = 11$$

(ii) when $y = 8$,

$$\Rightarrow 8 = 3(x) + 5$$

$$\Rightarrow \ 8-5 = 3x$$
$$\Rightarrow \ 3 = 3x$$
$$\Rightarrow \ x = 1$$

Example 8.3

(a) Find the equation of a straight line that passes through the points (3, -2) and (8, -3).

Solution 8.3

Let $(x_1, y_1) = (3, -2)$ and $(x_2, y_2) = (8, -3)$

Therefore,

$$m = \frac{y_2 - y_1}{x_2 - x_1} = \frac{-3 - (-2)}{8 - 3} = \frac{-3 + 2}{5} = \frac{-1}{5}$$

The line has the slope $m = -\dfrac{1}{5}$ and passes through (3, -2).

Applying the slope-point straight line equation gives

$$y - k = m(x - h)$$

$$\Rightarrow \ y - (-2) = -\frac{1}{5}(x - 3)$$

$$\Rightarrow \ y + 2 = -\frac{1}{5}(x - 3)$$
$$\Rightarrow \ 5(y + 2) = -(x - 3)$$
$$\Rightarrow \ 5y + 10 = -x + 3$$
$$\Rightarrow \ 5y = -x + 3 - 10$$
$$\Rightarrow \ 5y = -x - 7$$

$$\Rightarrow \quad y = -\frac{1}{5}x - \frac{7}{5}$$

Example 8.4

A straight line is represented by the equation $2x - 3y = -7$. What is the gradient and the y-intercept?

Solution 8.4

Notice that the equation of the straight line $2x - 3y = -7$ is in the general linear form. To find the gradient and y-intercept, rearrange the equation to the slope-intercept form, $y = mx + c$ where m, that is, the number in front of x, gives the gradient and the constant term c gives the y-intercept.

Now

$$2x - 3y = -7$$
$$\Rightarrow \quad -3y = -2x - 7$$

Multiplying by -1 gives

$$3y = 2x + 7$$

Dividing by 3 gives

$$y = \frac{2x + 7}{3}$$

$$y = \frac{2}{3}x + \frac{7}{3}$$

Therefore,

$$\text{slope } m = \frac{2}{3} \text{ and the } y\text{-intercept, } c = \frac{7}{3}$$

In summary,

(a) If a straight line passes through two points $A(x_1, y_1)$ and $B(x_2, y_2)$ then the slope or gradient of the line is given by:

$$m = \frac{y_2 - y_1}{x_2 - x_1}$$

(b) The general equation of a straight line may be expressed in three forms:

    (i) The **point-slope** form given by $y - k = m(x - h)$

    (ii) The **slope-intercept** form given by $y = mx + c$

    (iii) The **general linear** form given by $ax + by + c = 0$

8.4 Eyeballing the best-fit straight line

With this method of fitting—the best-fit straight line—an individual looks at the scatter plot and *simply* tries to draw the best straight line through all the data points such that the spread of the data points on either side of the fitted line is as equal as possible. Having fitted the straight line to the scattered data points, an equation representing the line is then determined. This is readily done by selecting any two points that lie on the line and using them to find the slope m and the y-intercept c. The disadvantage of this method is that the fitted line is dependent on the observer drawing the line.

8.5 Eyeballing Using the Centroid

With this method, individual judgment is still used in fitting the straight line between the data points of the scatter plot, but the line must pass through the **centroid**. The centroid of the data points is denoted by $(\bar{x}, \bar{y})$, where $\bar{x}$ is the average or mean of all the x coordinates of the data points, and $\bar{y}$ is the average or mean of all the y coordinates of the data points. Again, once the straight line is drawn, any two points that lie on the line may be selected to calculate the slope and y-intercept of the fitted line.

8.6 The Method of Least Squares

The method of least squares is used to insert the best-fit straight line among the scatter plot in an objective manner, thereby removing individual judgment. The line fitted to the points using this method is called the **least square straight line**.

To understand the concept behind the least square straight line, consider figure 8.4, which shows seven points and the best-fit straight line AB. For any given x value, say x_2, the actual y value is y_2, but the corresponding y value as determined (predicted) from the straight line AB is $\hat{y}_2$. As shown in figure 8.4, the difference between the actual value y_2 and the predicted value $\hat{y}_2$, that is, $(y_2 - \hat{y}_2)$, is referred to as the error or deviation and is denoted by e_2.

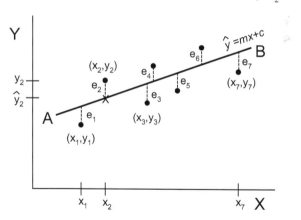

**Fig. 8.4 Least Square Straight Line
Showing Errors of Prediction**

Since the fitted line in general tries to divide the scatter plot into two equal sections, then most of the actual data points will not lie on the straight line itself. And so most of the predictions from the fitted straight line would generate deviations or errors—e_1 to e_7—as shown in figure 8.4. The method of least squares says that the best-fitting straight line is the line that minimizes the sum of the squared errors. That is, $e_1^2 + e_2^2 + e_3^2 + e_4^2 + e_5^2 + e_6^2 + e_7^2$ is a minimum.

8.7 Determining the Constants of the Least Square Line

Similar to any straight line, the least square line has the general equation $y=mx+c$, where m is the slope of the line and c is the y-intercept.

The slope of the least square line is given by

$$m = \frac{SSxy}{SSxx}$$

where

$$SSxy = \Sigma xy - \frac{(\Sigma x)(\Sigma y)}{n}$$

$$SSxx = \Sigma x^2 - \frac{(\Sigma x)^2}{n}$$

n = number of measurements

and the y-intercept c of the straight line is given by

$$c = \bar{y} - m\bar{x}$$

$\bar{y} = \frac{\Sigma y}{n}$, the mean of the y-coordinate values and

$\bar{x} = \frac{\Sigma x}{n}$, the mean of the x-coordinate values.

8.8 Interpreting the Least Square (Regression) Straight Line

Having determined the equation defining the regression straight line, it may be used for predicting or estimating one variable value given the other variable value. When the variable being predicted lies within the range of the collected data, it is referred to as **linear interpolation**. If the regression line is extended beyond the range of the collected data in order to predict a variable value, it is referred to as **linear extrapolation**.

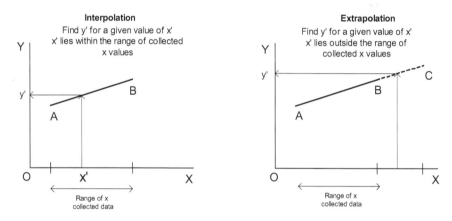

Fig. 8.5 Interpolation Compared with Extrapolation

Figure 8.5 shows graphically the difference between interpolation and extrapolation for a linear regression line AB. The straight line AB may be used to predict or estimate a y value given an x value. For the given x value, draw a vertical line until it intersects with the line AB. From the point of intersection, draw a horizontal line to cut the y-axis. Where it cuts the y-axis, read off the predicted value y'. If the predicted value of y' is for a value of x included between collected values of x, the process is called linear interpolation. If the predicted value of y is for a value of x that lies outside the collected values of x, the process is called linear extrapolation. Note in this case that to estimate the corresponding value y' for a given value x', the line AB is extended (extrapolated) from B to C. Extrapolation must be done with caution as we are essentially dealing with the unknown, meaning that we are not sure if linear behaviour between two variables continue to exist outside the range of the empirical data collected.

Example 8.4

The table below shows the heights to the nearest centimetre and the weight to the nearest kilogram of 16 new female students enrolling in a secretarial program.

Height x(cm)	Weight y (kg)
155	47
160	52
163	53
163	55
165	54
165	58
167	57
168	60
168	62
168	63
170	63
170	65
170	68
173	70
175	72
180	77
$\Sigma x = 2680$	$\Sigma y = 976$

(a) Draw a scatter plot diagram of the data.
(b) Use the eyeballing method to fit the best straight line to the data.
(c) Determine the equation of the straight line constructed in (b).

(d) By first calculating the coordinates of the centroid for the data, construct the best-fit straight line.

(e) Determine the equation of the straight line constructed in (d).

(f) Construct the least square straight line for the data.

Solution 8.4

(a) The scatter plot is obtained by plotting the points (155, 47), (160, 52), (163, 53) . . . (180, 77). The scatter plot is shown in figure 8.6.

Fig. 8.6 Scatter Plot of Height vs Weight

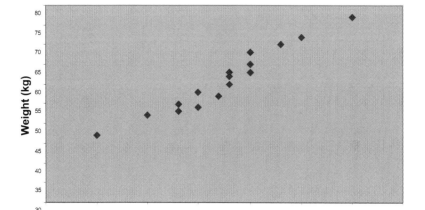

(b) Using the eyeballing method, a straight line approximating the data is drawn in figure 8.7. This line is one of many possible lines that could have been drawn, as it is dependent on the individual who draws in the line.

Fig. 8.7 Fitted Straight Line- Eye Balling Method

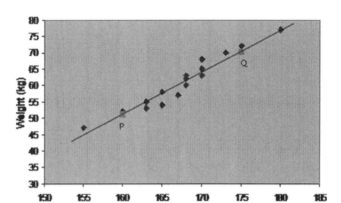

(c) To determine the equation of the line, first choose two points that lie on the straight line. For example, point P with coordinates (160, 51) and point Q with coordinates (175, 70) both read from the graph. Using these coordinates, the slope of the line is calculated.

$$\text{Thus, slope } m = \frac{y_2 - y_1}{x_2 - x_1} = \frac{70 - 51}{175 - 160} = \frac{19}{15}$$

$$\Rightarrow m = 1.27$$

Knowing the slope m= 1.27, and using the point P(160, 51), the equation of the line can now be found using the point-slope method, applying

$$y - k = m(x - h)$$

$$\Rightarrow \quad y - 51 = 1.27(x - 160)$$

$$\Rightarrow \quad y - 51 = 1.27x - 203.2$$

$$\Rightarrow \quad y = 1.27x - 203.2 + 51$$

$$\Rightarrow \quad y = 1.27x - 152.2$$

Thus, the equation of the eyeballed fitted line is $y = 1.27x - 152.2$.

(d) The centroid is denoted by coordinates $(\bar{x}, \bar{y})$ where $\bar{x} = \dfrac{\Sigma x}{n}$ and $\bar{y} = \dfrac{\Sigma y}{n}$

The number of data points is 16.

$$\Rightarrow \quad n=16$$

From the above table, $\bar{x} = \dfrac{\Sigma x}{n} = \dfrac{2680}{16} = 167.5$

and $\bar{y} = \dfrac{\Sigma y}{n} = \dfrac{976}{16} = 61$.

Therefore, the centroid is (167.5, 61). The centroid is now plotted on the scatter plot diagram and the best-fit line is constructed such that it must pass through the centroid. The orientation of this line through the centroid is also a matter of judgment on the part of the individual who draws the line.

Fig. 8.8 Fitted Straight Line- Method of Centroid

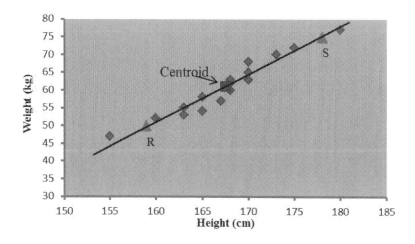

(c) To determine the equation of the line that passes through the centroid, we again proceed by first choosing two points that lie on the straight line. For example, point R with coordinates (159, 50) and point S with coordinates (178, 75) are both read from the graph. Using these coordinates, the slope of the line is calculated.

$$\text{Thus, } m = \frac{y_2 - y_1}{x_2 - x_1} = \frac{75 - 50}{178 - 159} = \frac{25}{19}$$

$$\Rightarrow \quad m = 1.32$$

Knowing the slope $m = 1.32$ and using the point R(159, 50), the equation of the line can now be found using the point-slope method, applying

$$y - k = m(x - h)$$

$$\Rightarrow \quad y - 50 = 1.32(x - 159)$$
$$\Rightarrow \quad y - 50 = 1.32x - 209.9$$
$$\Rightarrow \quad y = 1.32x - 209.9 + 50$$
$$\Rightarrow \quad y = 1.32x - 159.9$$

Thus, the equation of the straight line passing through the centroid is $y = 1.32x - 159.9$.

(e) The required least square straight line is given by

$$y = mx + c$$

where

$$m = \frac{SSxy}{SSxx}$$

$$SSxy = \Sigma xy - \frac{(\Sigma x)(\Sigma y)}{n}$$

$$SSxx = \Sigma x^2 - \frac{(\Sigma x)^2}{n}$$

n = number of measurements

and c is given by

$$c = \bar{y} - m\bar{x}$$

where

$$\bar{y} = \frac{\Sigma y}{n} \quad \text{and} \quad \bar{x} = \frac{\Sigma x}{n}$$

The calculations involved in computing the sums are arranged in the table below.

x	y	x^2	xy
155	47	24025	7285
160	52	25600	8320
163	53	26569	8639
163	55	26569	8965
165	54	27225	8910
165	58	27225	9570
167	57	27889	9519
168	60	28224	10080
168	62	28224	10416
168	63	28224	10584
170	63	28900	10710
170	65	28900	11050
170	68	28900	11560
173	70	29929	12110
175	72	30625	12600
180	77	32400	13860
Σx=2680	Σy=976	Σx^2=449428	Σxy=164178

Since $SSxy = \Sigma xy - \dfrac{(\Sigma x)(\Sigma y)}{n}$,

$$SSxy = 164,178 - \dfrac{(2680)(976)}{16}$$

$$= 164,178 - \dfrac{2,615,680}{16}$$

$$= 164,178 - 163,480$$

$$= 698$$

and $SSxx = \Sigma x^2 - \dfrac{(\Sigma x)^2}{n}$

$$\Rightarrow \quad SSxx = 449,428 - \dfrac{(2680)^2}{16}$$

$$= 449,428 - \dfrac{7,182,400}{16}$$

$$= 449,428 - 448,900$$

$$= 528$$

Therefore, $\quad m = \dfrac{698}{528} = 1.3$

Now $\quad \bar{y} = \dfrac{\Sigma y}{n}$

$$= \dfrac{976}{16} = 61$$

and

$$\overline{x} = \frac{\Sigma x}{n}$$

$$= \frac{2680}{16} = 167.5$$

Therefore, $c = 61 - (1.32)(167.5)$

$$= 61 - 221.1$$

$$= -160.1$$

Thus, the equation of the least square regression line is $y = 1.32x - 160.1$.

The graph of the regression straight line is shown in figure 8.9. The value of the slope 1.32 is interpreted to mean that for every unit increase in height of a person, her weight increases by 1.32 units. The intercept value –160.1 can only be interpreted practically only if the value $x=0$ is meaningful. In this problem, $x=0$ means that a person has no height, which clearly does not make sense, so the value of $c = 160.1$ has no practical interpretation.

Fig. 8.9 Least Squares Straight Line- Linear Regression

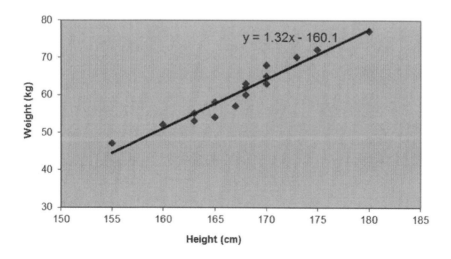

Consider the straight line equations obtained by the various methods:

Eyeballing only	$y = 1.27x - 152.2$
Centroid and eyeballing	$y = 1.32x - 159.9$
Least squares	$y = 1.32x - 160.1$

The equation of the least squares line is the most accurate description of the data points given in the problem, as the method of determining this line is objective and rigorous. The equation of the centroid-derived line is very close (in this case) to the least squares line, whereas the line derived strictly by the free hand approach, which is very subjective in fitting among the data points, is the least accurate. It should also be noted that the first two methods require one to first draw the straight line and then determine its equation, whereas with the least square method, the equation of the line is determine strictly from the data points given.

If it was now required to predict the weight of a girl whose height is 172 cm using the regression equation, we would substitute $x = 172$ into the equation and solve for y.

$$y = 1.32(172) - 160.1$$
$$y = 227.04 - 160.1$$
$$y = 66.94cm$$

Statistical Terms

linear regression. A statistical technique that is used to generate a best fit straight line for a scatter data plot

gradient (slope). A measure of the steepness of a straight line

y-intercept. The point where a straight cuts the y-axis

centroid. The coordinates of a scatter plot denoted by $(\bar{x}, \bar{y})$ where $\bar{x}$ is the average of all the x coordinates of all the data points and $\bar{y}$ is the average of all the y coordinates of all the data points.

INDEX

Range 72, 89, 90, 119, 133
Raw data 4, 11, 12, 22, 24, 42, 75, 76
Relative cumulative frequency 57, 70
Relative frequency 19, 21, 56, 57, 58, 61, 65
Relative standing 103, 104
Representative sample 7

S

Sample 6, 7, 11, 23, 28, 56, 57, 73, 88
Sample data 53
Scatter plot 112, 113, 115, 121, 130, 131, 137
Skewed frequency distribution 45, 47
Slope 122, 123, 125, 131, 132, 136, 138, 143
Slope-intercept 126
Smoothed frequency polygon 43, 47
Standard deviation 92, 93, 95, 102, 109, 110
Standardized normal curve 110

U

Upper class boundary 30, 32
Upper class limit 29, 32

V

Variable 4, 6, 11, 13, 16, 23, 24, 26, 34, 73, 115, 124, 133
    continuous 24, 32
    discrete 24, 32
    normal 109, 120

qualitative 4, 11
quantitative 4, 11
standardized normal 110
Variance 93, 120

Y

y- intercept 125, 132, 143

Z

z- scores 104

Printed in Great Britain
by Amazon